# HIGH RISE AND FALL

## The Making of the European Real Estate Industry

*Andrea Carpenter*

Routledge
Taylor & Francis Group

LONDON AND NEW YORK

First published 2019
by Routledge
2 Park Square, Milton Park, Abingdon, Oxon OX14 4RN

and by Routledge
711 Third Avenue, New York, NY 10017

*Routledge is an imprint of the Taylor & Francis Group, an informa business*

*British Library Cataloguing in Publication Data*
A catalogue record for this book is available from the British Library

*Library of Congress Cataloging in Publication Data*
Names: Carpenter, Andrea, 1972- author.
Title: High rise and fall : the making of the European real estate industry /
Andrea Carpenter.
Description: Abingdon, Oxon ; New York, NY : Routledge, 2018. | Includes
bibliographical references.
Identifiers: LCCN 2018023662| ISBN 9781138612419 (hardback : alk. paper) |
ISBN 9781138612426 (pbk. : alk. paper) | ISBN 9780429465031 (ebook)
Subjects: LCSH: Real estate business--Europe.
Classification: LCC HD586 .
C37 2018 | DDC 333.33094--dc23LC record available at https://lccn.loc.gov/
2018023662

ISBN: 978-1-138-61241-9 (hbk)
ISBN: 978-1-138-61242-6 (pbk)
ISBN: 978-0-429-46503-1 (ebk)

Typeset in Bembo
by Taylor & Francis Books

MIX
Paper from
responsible sources
FSC
www.fsc.org   FSC™ C013985

Printed in the United Kingdom
by Henry Ling Limited

For Dad, who preferred his land without buildings.

# CONTENTS

| | |
|---|---|
| *Acknowledgements* | *x* |
| *Notes* | *xi* |
| *Abbreviations* | *xii* |
| *Foreword* | *xiv* |
| *Introduction* | *xvi* |
| 1 The Americans in Paris | 1 |
| 2 Money coming out of boxes | 11 |
| 3 For the love of property | 26 |
| 4 The return of the Europeans | 38 |
| 5 A rising risk in Europe | 50 |
| 6 A new relationship with debt | 57 |
| 7 Blurring the lines | 69 |
| 8 Crisis hits Germany | 80 |
| 9 The final party | 93 |
| 10 Credit crunch, and denial | 104 |
| 11 Life after Lehman | 116 |
| 12 Bailouts and workouts | 125 |
| 13 Revival and survival | 135 |
| *Bibliography* | *147* |
| *Index* | *165* |

# ACKNOWLEDGEMENTS

I would like to thank everyone who agreed to be interviewed as part of the research for this book. They very generously raked over good and bad memories of the period to help bring context and meaning to what was a whirlwind time in the industry.

My special thanks to Viktorija Grubesic, Stephen Rowe and Lucy Scott for their unending support and encouragement for what often appeared to be an unending project.

Thanks to Alex Catalano for her support, insights into the text and sharing with me her wealth of knowledge on the European property industry.

Many others have been a great help to me at various stages of this book, for which I am very grateful, including David Allen, Karyn Arthur, Alison Jones, Evelyne Nossing, Will Powell, Jeff Rupp and Jamie Vuong.

Finally, I would like to thank Ed Needle, Catherine Holdsworth and Lisa Sharp from Taylor & Francis for guiding this book through to publication.

# NOTES

**Currencies**: Figures have generally been left in the original currency that they were reported in at the time. For amounts in pre-euro currencies, a euro conversion has been included based on the fixed exchange rates at the time of adoption. Any other conversions were calculated to the same month and year as the information was published, using the historical rates on oanda.com.

**Company names:** Where possible, the text tries to reflect the names of the companies as the brands changed. There may be some slight timing differences to avoid changes in company names mid-chapter.

**Rounding of figures:** where research figures such as percentages for market growth or size have been used, they have in general been rounded up or down to the nearest whole number. This is for ease of reading. For figures that require a greater level of accuracy, such as yields, the decimal places have been left.

**Interviewees' job titles/references:** Often people work at different companies now than the time about which they are speaking. If this is the case, it mentions the companies at which they worked in that period. If that context is not relevant or they are still at the same company, then their current job title is given.

# ABBREVIATIONS

| | |
|---|---|
| 3D | three-dimensional |
| ABS | asset-backed securities |
| ACSI | Ahorro Corporación Soluciones Inmobiliarias |
| AIFMD | Alternative Investment Fund Managers Directive |
| AIM | Alternative Investment Market |
| AUM | assets under management |
| BaFin | Bundesanstalt für Finanzdienstleistungsaufsicht |
| CDC | Caisse des Dépôts |
| CDO | collateralised debt obligation |
| CDR | Consortium de Realisation |
| CHF | Swiss franc |
| CIC | China Investment Corporation |
| CMBS | commercial mortgage-backed securities |
| CNIT | Centre des Nouvelles Industries et Technologies |
| CSFB | Credit Suisse First Boston |
| DIY | do it yourself |
| DM | Deutsch mark |
| ECB | European Central Bank |
| ELoC | European Loan Conduit |
| EPAD | Établissement public pour l'aménagement de la région de la Défense |
| EPRA | European Public Real Estate Association |
| EUR | euro |
| FCP | Fonds Commun de Placement |
| FFr | French franc |
| GCI | Générale Continentale Investissements |
| GP | general partner |

| | |
|---|---|
| G-REIT | German real estate investment trust |
| HVB | HypoVereinsbank |
| INREV | European Association for Investors in Non-Listed Real Estate Vehicles |
| IPO | initial public offering |
| IRR | internal rate of return |
| L | lira |
| LP | limited partner |
| LTV | loan to value |
| M&A | mergers and acquisitions |
| MSREF | Morgan Stanley Real Estate Fund |
| NAMA | National Asset Management Agency |
| NAV | net asset value |
| NLG | Dutch guilder |
| NN | Nationale Nederleden |
| NPV | net present value |
| p | pence |
| Q&A | questions and answers |
| RBS | Royal Bank of Scotland |
| REIT | real estate investment trusts |
| RMBS | residential mortgage-backed securities |
| RTC | Resolution Trust Corporation |
| SAREB | Sociedad de Gestión de Activos Procedentes de la Reestructuración Bancaria |
| SIICs | Sociétés d'Investissements Immobiliers Cotées |
| SKR | Swedish krona |
| sq m | square metre(s) |
| ULI | Urban Land Institute |

# FOREWORD

When I came back to work after the Christmas holidays in early 2012, I was on the board of a company that - in anticipation of economic recovery - had more than one million square feet of speculative space under development in London, and there wasn't a tenant in sight.

We were also still in the midst of an economic crisis; each week, we were having calls to discuss how to manage the business if the euro collapsed. It wasn't until later in the year that European Central Bank president Mario Draghi's promise to do "whatever it takes" finally saved the single currency.

Post-global financial crisis, juggling these two different scenarios is what property has evolved into. On one hand, dealing with the very tangible reality of day-to-day real estate topics. On the other, trying to anticipate and factor in the risks of geopolitics and economics.

Coming into the industry now, this mix of challenges seems normal but it has been the result of a staggering pace of change in the last two decades that brought us to a new level of financial sophistication and transformed my working life.

From arriving as a graduate at Jones Lang Wootton (now JLL) through to my role now as CIO at Patrizia, every step in my career has been further into the world of international investing. These were interesting and exciting times, the arrival of the US opportunity funds and later the euro prompted a wholesale change in the way business was conducted globally.

I'm delighted that Andrea has now spent time capturing this transformation as the opening up of the European markets has been an amazing shift for property but one that we rarely stopped to consider or document. The culmination was the global financial crisis – which she has spent time analysing – and which caused the industry a great deal of pain. Now, post crisis it has resulted in a positive shift that brought about a much greater role for real assets in the financial markets.

I've known Andrea for most of her career in the real estate industry and think that she has a unique vantage point from which to tell this story. Starting as a journalist, she edited *EuroProperty* as the pan-European industry really started to form. Later at INREV, she was immersed in the world of non-listed real estate funds, one of the most dominant investing styles in Europe up until the crisis.

At each step in her career she has built up a network of trusted relationships with industry figures who have now shared their views in this book. In addition to those quoted, many others contributed and there are many interesting stories.

As well as gaining a better understanding of the roots of this modern industry and the path to the crisis, those reading this book will see the origins of the many companies that lead the industry today (and also the fate of some of those that got caught up in the crisis). Andrea's insights also help us to understand some of the risks for future cycles - in some cases, lessons we thought we had learnt such as the importance of liquidity, not being a forced seller and not over-stepping our relationship with debt.

Today, we are experiencing a new set of challenges. The influence of technology at all levels of society is impacting the way we communicate, live, work and play. In turn this is bringing major structural changes in our use and relationship with real estate which we need to understand to invest successfully. The next 20 years are likely to see even greater change than in the past and I believe it will be a fascinating time to be in this exciting industry. I hope you enjoy Andrea's book and take inspiration from it.

Anne Kavanagh
Chief Investment Officer (CIO) Patrizia July 2018

# INTRODUCTION

Shortly after the Lehman Brothers crash, I attended an event at which the presenter put up a chart showing the effect of debt on property performance. It was a graph plotting returns against rising levels of leverage, two curved lines one above and one below the x-axis that didn't quite mirror each other. Mathematically, what it was trying to convey was that in an up market debt was a helping hand, on the way down it was a punch in the face.

You'll know the graph I mean, or may well have studied it, readying yourself to enter an industry that, after the Lehman collapse, was defined by that chart's teachings. But for many in the room on that day – me included – it was the first time they'd properly focused on it. This was an industry that had gloried in a prolonged bull run and for those that followed into the business afterwards, the years before the crash became an increasingly less relevant time.

As new people enter this profession, even ten years on, the industry still lives with the scars of the global financial crisis. The markets might feel healthier, to the point that in early 2018 all talk is of later-cycle investing, but the crash is still omnipresent in the industry's views on taking risk, where and how it invests, and its future prospects relative to the other financial asset classes.

I wrote this book because I believe to have a better understanding of the industry is not just to look at the years of the crash but also the origins of the modern version of the business that you work in or study today.

That 15-year period in the run-up to the global financial crisis was not just another cycle but a complete reset for the property industry. It was a structural shift that took it from a local, small-scale industry contained within national boundaries, into the global, financially sophisticated billion-dollar empire-building business that it is today.

Understanding more about how this industry evolved in this modern era will not only give you a better insight into why the crash was so devastating, but why now – well into the next cycle – property is still learning to find its place in the wider financial landscape.

This story starts in the mid-1990s, the beginning of the cycle but also the period that marks the start of the modern industry. It looks at the influence of the arrival of US money and how it transported high-return investing and its grand macro-economic vision for property across the Atlantic. It also covers how Europe emerged from the end of a development-led crash in the early 1990s to create a league of increasingly financial pan-European players with deeper pockets and bigger ambitions.

Their back stories trace a path to the crash – and out the other side – from much simpler times through to scale and complexity that still presents the industry with challenges today. It's a European story but one which in an age of financial deregulation, increased capital and information, was part of a similar change happening globally. There is no more obvious sign of this today than the multi-cultural make-up of the money that owns property in Europe.

This is an industry where not much is written for the next generation that is not in textbooks or research papers. It can be easy to forget that in addition to the spreadsheets, human nature and behaviour – good and bad – often define many of the decisions behind the industry's evolution.

This book is based on more than 100 one-to-one interviews in person and by phone with those working in the industry during this period. They gave their time generously and were thoughtful and insightful as I pressed them to recast years of investing into an historical context.

In general, people were happy to talk on the record but for some topics those involved were reluctant, so I've had to lean more on my research and the non-attributable views of those working in the same sphere.

The story is also influenced by my time working in European property. I've been fortunate enough to witness much of its transformation from viewpoints that gave me access to those leading the change, and had the privilege to work in an international industry that spans so many cultures.

Andrea Carpenter, April 2018

# 1

# THE AMERICANS IN PARIS

When François Trausch tried to turn up for his first day in his new job in the summer of 1997, he had no idea that he was about to enter a market in mid-transformation.

As he headed to GE Capital's offices in Paris, his mobile phone rang. At the end of the line was his new boss, who diverted him instead to an anonymous building in central Paris.

When the lift door opened, Trausch stepped into a data room. Inside there were close to 100 people sifting through dozens of boxes, taking out paper files and manually inputting the data from them into computers. They were analysts like him – a few French, but largely a team that was flown in from the US.

Each file held details of a broken property loan stretching back as far as 1990. Setting to work, Trausch could see exactly how banks overextended: the lending terms, the poor decisions, the bad borrowers, all the steps that led to the systemic scale of poor lending. One file at a time, Trausch relived the worst French property crash in history.

The story told in those files from the early 1990s used to be the ultimate cautionary tale for the industry. Paris was a city that found itself at the tail end of a zealous period of overbuilding. All in all, from 1986 to 1991, more than 12 million sq m of buildings rose from the ground, expanding office space in the city by 40%.

Collectively, the efforts of developers in Paris were historic, city-making, even. At one point, Paris marched past London to become Europe's largest office market. A few years later, it surpassed even New York.

To have the Paris skyline packed with construction cranes in the late 1980s felt good for everyone; visible signs of the booming French economy, the property industry a driving force in progress, backing city growth to the hilt. Then, the world tuned into its televisions to watch live as fighter jets returned from sorties in Iraq, marking the beginning of the first Gulf War in 1991. The mood changed, as did the direction of the economy. France quickly slumped into recession and the same cranes soon became a shameful display of excess, a daily reminder of the hubris of business, and of the property industry.

The fallout for the property industry was devastating, but it could in no way be described as fast. No cranes creaked to a halt mid-development, no rents went into free fall overnight. This was an old-style property crash, and the end was dismal and protracted as the market dismantled itself over four long years.

At first, it was almost as if no one in the industry noticed. Developers finished up buildings into an expanding vacuum of demand from businesses. Others started new projects as the banks stayed firm on agreements to lend them the money. But not everyone remained so upbeat. With the growing burden of oversupply, investor confidence slipped, and values fell as the declining rents agreed for new space only confirmed the weakening market.

By 1995, the Paris property market had inched itself down to the bottom. Buildings were worth less than half of the FFr102,000 per sq m five years before. Rents too dropped from a peak of FFr5,000 in 1990 to just FFr3,000.[1]

The same year the entire French property market bought and sold just FFr3 billion (EUR457 million) of property. Today, a single company somewhere in the world completes a deal of that size – or more – each and every week.

This stumbling pattern of decline was the quintessential property crash. This was how it worked in those days before the industry became so much more embedded into the financial landscape. Property was a business, local and clubby, slightly set back from the rest of the financial markets, the economy an engine running in the background. Property investment markets in Europe were all domestic, propped up by local money. There was no glut of global money queuing up to compete with the home players or plug the gap left by the locals if they got burnt in the downcycle.

For institutional investors – the pension funds and insurance companies – property was just property. That department down the end of the corridor, the one with the big headcount for the relatively small amount of money to invest compared to bonds and equities. That large team, however, earned its keep as when it came to performance, property always acted differently; a dependable counter-balance to the risk of equities, usually earning a bit more yield than bonds.

Today, as a truly global and financial industry, property works on much bigger levers. It is interest rates that steer the fortunes of property, acting as the valve to control how much money flows into the European industry from all parts of the globe. Its prospects are now as much about geopolitics as they are geography.

In today's prolonged low interest rate environment, property is massively in favour. It has a new, more dynamic role in the portfolio alongside bonds, equities and other alternatives. Now, they call it a good fixed-income alternative in a world of low-yielding bonds.

This new set of clothes has come at a price. With the global financial crisis in 2008, the inevitable crash for property was not the usual slow disassembling of the markets. Instead, property fell off the cliff at exactly the same time as the rest of the asset classes. Unprecedented crash or not, property was hooked into the deepest reaches of the global financial system.

This makes the Paris crash sound nostalgic. It is unlikely that the industry will ever see a crash like that again, one where property was its own cycle, tangled up in its isolated out-of-kilter mess of demand and supply, and with only itself to blame when it all went wrong.

But that doesn't mean to say Paris should be forgotten. In actual fact, it's the first reference point of the modern market. As recent as the scars of the global financial crash still feel, the industry is where it is today not because of the global meltdown but because it changed so utterly in the preceding years. And all that started with Paris.

The Paris crash was so off-the-charts bad that the industry has never been the same again. In a business that is unfailingly cyclical, with generations of players programmed to repeat a suite of the same mistakes, Paris was the kink in the system that transformed the industry into the financial giant it is today.

## Getting Coeur Défense off the ground

There were souvenirs from La Défense all around the room at Paul Raingold's offices. Fixed to one wall were two giant pictorial maps of the business district to the north-west of Paris: a cheerful poster, with colourful 3D renditions of each building. The second, a sober out-sized version of today's visitor's map. In another part of the room, a small architect's model encased in glass was perching awkwardly on the rim of the lower shelf of a hostess trolley. The gleaming high-rise tower trapped inside, a planned refurbishment on the second map.

For a man that is reluctant to be reminded about Coeur Défense, the lowest point of his career, Raingold finds himself in plain sight of this misfortune. Raingold is a successful man. For over 40 years, he has added and updated buildings to those maps. Not getting to build Coeur Défense hardly put a dent in his lifetime achievements, but he understandably remains unconvinced. "It's something we have to live with," he said. "We all made losses but c'est la vie, we did our best."

It is difficult to believe that this doorstop of a building almost never happened. Coeur Défense is by no means the tallest building in Paris, and not the most elegant by a substantial margin. Its two 39-storey towers sit above three eight-storey groundscrapers, a bottom-heavy shape that had given it an air of sinking middle-age spread. Yet, it has always held a commanding presence at La Défense. In Paris, this is the place to which urban overachievers are banished, far far away from the city's fairy-tale centre. There are at least 30 buildings over 100 metres high. Among them, Tour First, the country's tallest at 231 metres. Grouped together in one tribe, they are impressive, but in reality, they all defer to Coeur Défense. It's not just that they lack heft, they also lack history. When Coeur Défense was finally built, it became Europe's largest single office building and still today, this heavyweight retains the title.

Raingold, with his development and investment company Générale Continentale Investissements (GCI), was on the front foot during this massive expansion of Paris in the late 1980s. Along with a partner, he had just sold six of a series of seven low-rise buildings each side of La Grande Arche, the triumphant arch that sits at the head of the La Défense estate. Those sales gave Raingold the equity for his greatest prize, Coeur Défense.

As a project, it was a gift, even Raingold acknowledged that. With La Défense centrally masterplanned and promoted by government quango Établissement public pour l'aménagement de la région de la Défense (EPAD), Coeur Défense came oven-ready. "It was 160,000 square metres of offices and it had planning permission, which was very unusual," he said.

His company was part of an unwieldy group, its 11 investors making up the only consortium that bid for the site. Along with GCI, the group contained two other developers, one insurance company and seven banks. Unusually, the banks did not just agree to lend money but also took major stakes in the deal. The upgrade from lender to equity provider was already a sure sign that things were overheating in the city.

The deal included Tour Esso, the existing building on the site. Itself an artefact, it was nestled into a generous 2 hectare piece of land bought by oil company Esso back in the early 1960s. As the first office to have been built at La Défense, Esso's first-mover advantage was lonely. For the next five years, its 1,500 employees practically trudged through mud each day to reach their desks. With only the newly built CNIT (Centre des Nouvelles Industries et Technologies) exhibition centre as a neighbour, and no public transport, Esso was ever-patient as the wasteland that surrounded it slowly transformed into a corporate city. At 11 storeys, Tour Esso was the site's first "high-rise" building, but now in Coeur Défense's new generation of skyscrapers, it was considered a waste of good airspace.

The group paid FFr2.9 billion (EUR442 million) to buy Tour Esso, the site and the development rights for the new tower.[2] (A figure that amounted to the same volume as the entire industry bought in that record low five years later.) There was just one very large problem. If Coeur Défense had risen out of the ground, it would have sat on top of the other 4 million sq m of office space that already lay vacant in the city, most of it also newly built. One-tenth of all offices in Paris were completely empty, equivalent to another 25 record-breaking Coeur Défenses.

This growth in Paris was impressive, particularly as it was a defensive measure. Paris had been forced to soften its usually tight-fisted stance on *agréments*, the planning permissions needed for buildings over 1,000 sq m. In its race to attract major corporations, it was being sideswiped by the competition and – it couldn't be worse – by London.

From across the Channel, Canary Wharf was winking at them. This was Prime Minister Margaret Thatcher's era. She placed no restrictions on East London's new masterplanned business district. Paris needed to retaliate.

Raingold and his fellow developers in Paris were the beneficiaries. From the mid-1980s, the *agréments* were lifted, and construction began. By the time the gate was lowered again in 1990, new buildings peaked at 2.5 million sq m a year.[3] This was at a rate equivalent to adding 15 Coeur Défenses every 12 months. Most of this space was speculative – built with no guarantee of tenants – and the over-whelming proportion was in La Défense.

Alongside the ease of development, a second factor was in play: the banks. Bank financing was essential to get major projects such as Coeur Défense off the ground, and this development boom had corresponded with a time when bank lending – on favourable terms – was more than readily available.

In the three years to 1990, the debt outstanding to developers soared to a peak of FFr173 billion (EUR26.4 billion), more than ten times the average over the previous decade.[4] In 1990 alone – the year before the consortium bought Coeur Défense – lending to developers added a further FFr35 billion (EUR 5.34 billion).[5] Overall property lending, not just to developers, rose to a peak of FFr375 billion (EUR57.17 billion) in 1993.[6]

The figure was so high because banks were not just lending to developers. For a large part, their clients were *marchands de biens*, or traders. These unregulated, casual individuals often turned properties at a ridiculous pace, buying and selling them on for big profits within a matter of months (breakneck speed for bricks and mortar).

These traders had one major advantage. As long as they sold the building within four years, they weren't required to pay an onerous 20% transfer tax on the sales price.

In the rising market, this propelled their rate of trading, leaving no time to make changes to the buildings and tenants to add value and justify the higher sales price. Instead, they traded on the positive sentiment in the market, cashing in on the rise in value of the building. With investment demand pushing prices up beyond the fundamentals, these *marchands de biens* sold buildings plus hot air, and that air quickly filled a classic property bubble.

## Back in the data room

When Trausch entered that data room in Paris following the crash, it was clear amid the papers, files and spreadsheets, that nothing about this looked like a regular property deal. Trausch was working for one of the new US investors in town, and he soon found himself steeped in a brand-new way of thinking about property. "We all went to the 'same school of the data room'," he said. "The new genera-tion of real estate in Europe, and in France, had all gone through these data room experiences. It was like a boot camp; an experience which people remember many years later."

Trausch was helping his American employer prepare its bid for the FFr1 billion (EUR152 million) portfolio of broken property loans locked down in those files. They planned to buy the loans at a fraction of their original cost and work them out, negotiating new payments from defaulting owners or repossessing the property.

In a moribund market, the domestic players were perplexed as to why this was happening. "It was very new in Europe and so people had no idea that values would recover," Trausch said. "Nobody believed in it. The only ones who believed in it were the American firms."

It was this faith that property markets always rebounded which underscored this new American style of investing. With that knowledge, American investors hadn't joined the crowds when the markets were booming. Instead, they set foot in Paris just as property hit its darkest moment. The crash pushed the city to the edge, and while that was painful for the locals, what the Americans saw – the mess, dysfunction and pain – was the perfect starting point.

The Americans were high-return investors, and their investing style was founded on unearthing value in distressed markets. In Paris, at the lowest point, they brought cash to capital-starved markets. When property rebounded, which they knew it would, the potential for returns over a short timeframe was stratospheric.

Landing in France was just the start. Buying up non-performing loans in Paris was the first demonstration of the transformative power of this type of capital. The American investors went on to spend EUR 90 billion across the continent over the next decade.

Just as a powerhouse of investing, the Americans left an indelible mark, but the lasting legacy continues to be their methods. They introduced high-risk investing to Europe on an unprecedented scale – starting in the data rooms in Paris – bringing with it a new mindset that influenced a next generation of pan-European players.

Today, investing anti-cyclically is standard but to do that 20 years ago on such a scale and so systematically was groundbreaking. It is now unthinkable that there used to be periods in the market when there were no buyers. Today, high-return buyers are always poised, nimble and efficiently moving around the world's markets anticipating the distress to come at the bottom of the cycle. "In the scheme of things," Trausch told me, "even though the amounts were much smaller than today, it was a bold move."

## Denial in the banking markets

It was the stink of the debt that finally put the American investors onto the scent in Paris. The city's property market had transformed into a textbook example of the pernicious nature of debt in property.

Excessive lending is nothing new. Debt has always had the power to be destructive on a systemic basis – not just for the property industry or banking, but for the whole economy. In a downturn, an overload of bad property loans will suffocate a bank's ability to lend across all industries. Debt has been, and always will be, a driving factor in property crashes, and while Paris was no exception, it was a rare example of being on a scale to have worried government.

In Paris, the banks were another willing player in a conspiracy of containment. As the market slowed and loans were breached, they held the line as values dropped and borrowers failed to pay. The silence left them with the bad debt on their books. Besides, to them, it made no sense to sell at lower prices. Instead, they were secret eaters, accustomed to binging but cunning enough to keep that shame private.

It had always been this way. In the very worst of times, banks may have offloaded something considered completely non-core, but if it was important to the bank strategically, or part of a long-term client relationship, it was buried, however bad.

But this time the Paris crisis was so big, that it left even the banking industry shifting uncomfortably in its seat. These were no longer problems that could be kept within the family. Across the board, banks' loan books were devastated by failing loans. The scale of the excessive lending was threatening the very existence of the banks, as write-offs wiped out any provisions already set aside, as well as the banks' profits. With many of the banks partly state-owned, the French government had no choice but to intervene.

Top of the pile was Crédit Lyonnais. It was Europe's largest financial institution, and acted very much in line with that status. All through the 1980s, it bought up rival international banks and expanded its lending across a host of industries. For a while it stumbled into the movie business when it became the default owner of the MGM-Pathé film studios.[7]

In deep trouble in the early 1990s, it further destabilised itself by trying to set aside massive provisions to cover its defaulting loans. By 1994, the French government stepped in with a bailout, and a rescue plan that saw it take on US$27 billion of defaulting loans, transferring them to a new unit called Consortium de Réalisation (CDR).

CDR included $8.5 billion in property loans, which was by all accounts most of Crédit Lyonnais' property-lending portfolio.[8] The bailout erased the mess of the past, leaving Crédit Lyonnais with a healthier balance sheet from which to begin new property lending. Meanwhile, needing to recoup some losses, CDR became a discount store for French property.

It was a sensible enough plan but it was also a shocking move for the market. Once CDR had acknowledged its non-performing loans, the curtain had lifted and there was pressure on others to do the same. CDR's actions threatened a wholesale unveiling of the banking sector's secrets. In addition, there were fears that the flood of sales from CDR would "find the floor" for pricing, marking a new low point from which sales in the market would begin.

The pressure the French government placed on banks to sell their non-performing loans was a watershed. Paris was a dangerous reminder of the real risks that property lending could represent on a bank's balance sheet. It acted as a release valve for the stultified market, at whatever cost to the property industry.

As a developer with seven banks invested in and lending on his prize project, none of this was at all helpful to Raingold. As the poison seeped deeper into banks' loans books, their ability to partner on Coeur Défense slowly dwindled. "Progressively as the problems of the banks grew, they saw more explicitly other projects in the crisis going down," he said. "They started to question whether such a difficult time was the right time to build on spec."

It was over for Raingold and his partners. The group held out until 1997 and just about managed to knock down Tour Esso, leaving a muddy wasteland of a hole where Coeur Défense should have stood. "Everyone says it's location, location, location," said Raingold. "But it's really timing."

## The Americans in Paris

The chasm between the French and the American investors finally broke open at the Urban Land Institute's (ULI) European annual conference in Paris in 1997. The event was set up to cultivate business relationships for the growing number of Americans in town. Instead, a fissure splintered right across the conference floor.

Alec Emmott, then working for Société Foncière Lyonnaise (SFL), witnessed the exchanges first hand. "One side of the room had people talking about returns and meaning unleveraged, all equity returns of 5–6%, which it had always been for the last 20 years," he said. "And, on the other side of the room, you had these people with blood dripping off their fangs, talking about rates of return of 20%. The two shouted at each other: 'That's not possible. That's not possible. That's not possible'."

Gothic horror aside, this was serious. It was a clear message from the Americans to the French: there were no rules for how property investment was done in France, only convention. The trouble may have started with the developers and traders, but now it was reverberating through the industry right down to the long-term institutional investors.

Even as the conference met, the American financial giant GE Capital was in the final stages of its latest deal, the one on which Trausch was working. It was about to buy a portfolio of non-performing property loans with a book value of FFr1 billion (EUR152 million). It would pay an amount equivalent to around one-third of that.[9]

Henri Alster, who moderated the day's conference proceedings for[10] ULI, overheard a conversation between a US opportunity fund manager, and the head of a now defunct French bank. "The head of the bank said 'but you don't understand'. He was trying to explain to him 'I make the decisions', and the US private equity manager said 'No, those days are over. This is gone. This is finished. *You* don't understand'."

For the French, this type of interruption was visceral, an unwarranted punch in the stomach. American capital sliced right through the status quo, and they were losing control of their own market. The bedrock of the French property industry may have existed on the traditional 5% or 6% returns but the presence of capital with much higher stakes was starting to change the landscape.

French pension funds and insurance companies had always led their own market, and they looked at property more as a way of preserving the value of money rather than making it. It was an inflation hedge; investment in property generally outpaced the rate of inflation at least maintaining the spending power of that capital. It suited French institutions to be long-term holders with a focus on income. This was the template for how all European pension funds and institutions invested. Slow and steady was the European way.

In the bad times, the French institutions were used to toeing the line along with everyone else. A consultant in the French market spoke to the boss of a British insurer based in Paris at the time, who said, "Do you want me to give you the values French valuers give us, or do you want me to tell you what we think the buildings are really worth, which is 20% to 30% below those values?"

Now, the Americans were jettisoning that way of life by the simple act of being prepared to buy at the right price, the market price. "Property is always for sale, it's just a question of pricing," said Emmott. "And if you don't believe it, wait until there is a bit of blood on the streets and then see what your property's worth. It's worth what [the Americans] give you for it. The French market place understood that life had changed."

In the end, it took just five deals to change the market in Europe. Between December 1995 and February 1997, American buyers bought around FFr10.2 billion (EUR2.15 billion) of non-performing loans. Goldman Sachs' Whitehall fund led the pack, picking up portfolios with a book value of FFr8.3 billion (EUR1.75 billion). GE Capital and Lehman Brothers took the remainder.

It was expected to be a much deeper market but the early sales to the Americans did indeed set the floor for prices, allowing realistic buying and selling to once again begin. This accelerated the recovery in Paris but it made the prices of later distressed portfolios higher, and less profitable. That general lift in values across the market also lightened the pressure on the banks to sell, with banks more than happy to retreat.

Richard Georgi, then working for Goldman Sachs, likened this process to putting lubrication in a seized engine. The market stagnates because potential buyers cannot justify the higher prices expected from sellers who refuse to accept that values have dropped. "And then something breaks the log jam, and all of a sudden they have no choice but to sell. Then, that creates this virtual cycle of improved economics and prices rise. So, it's not a coincidence that when we buy these non-performing loans they always perform better than we expect, because we are actually part of the clean-up process."

Clean-up is probably right. These portfolios were pretty nasty. Often just broken toys that the traders hadn't been able to flip quickly enough. They were small, run-down offices in off-pitch locations if they were very lucky, but more likely failed projects on the Côte d'Azur, grain silos or dodgy nightclubs and racing stables. "Property was just an underlying asset," Trausch said. "And, obviously, the property underlying the non-performing loans wasn't pretty, the only reason was money. We bought it for 30 cents on the dollar."

This reflected that underlying faith that the Americans had in market cycles. Being agnostic about the bricks and mortar ended up being the route to find greater riches later. "The money wasn't made in high-profile properties that partners were expecting. It ended up being a lot of low-quality assets that we [first] bought in France. We ultimately ended up buying some very nice real estate," Georgi said.

That said, it's still unclear if these loans made the investment banks any money. The first portfolio bought by Lehman Brothers had a book value of FFr875 million

(EUR 133 million), but was sold to the bank at "20-something" cents in the dollar, so around one-fifth of the estimated worth of the property. Keith Breslauer, who led the deal for Lehman, agreed that the early portfolios at least had the wind of the market behind them. "I think we underwrote it around 19% and probably would have made a 13% on the original assumptions but because the market did so well, we did much better, probably 30% plus."

Others said that French law made it difficult to wrestle ownership or settlements from borrowers, which meant that the workouts – extracting the value from the loans – was protracted and costly, quickly wiping out potential profits.

If these deals did lose the Americans money, that outcome has ended up being a footnote in the history of their arrival in Europe. The legacy is the intervention in Paris at that time, changing the course of the market, and eventually the industry.

Without American capital, the speed and pace of the recovery would have been altogether different. There was no other capital around to accelerate the change, and there were no other buyers for the bad debt. Among European investors, there was very little understanding of how to buy debt, or how to manage the workout to uncover the value. And, in reality the gap was bigger than just that missing skill set. This type of large-scale, higher-return investing was not even being considered by European investors; it was within neither their capability nor their appetite.

It was this gap that had presented the next opportunity for the Americans. Through buying non-performing loans, Paris had shown them that the concept of high-return investing travelled. Now, they had the rest of Europe to explore.

# 2

# MONEY COMING OUT OF BOXES

The fact that Léon Bressler took the time out of his busy agenda to stare into that giant pit that was Coeur Défense in late 1997 was a sure sign that recovery in Paris was finally underway. Surrounded by the concrete and corners of La Défense's stark modernist architecture, he surveyed the wasteland left by the partial excavation. "It really was a physical as well as a financial hole," he said.

Back then, Bressler was CEO of Unibail, a small but tenacious property company, righting itself after the crisis. These were early days in the recovery and despite Bressler's deal being backed by international investors, it was still the equivalent of a snake trying to swallow a goat. But Bressler was undeterred, so as the rest of Paris rolled towards its New Year celebrations, he signed up to take on the speculative development of Coeur Défense.

Construction started in 1998 and it became a reference point in how to catch a cycle. Bressler had the timing that had escaped Raingold. "People more or less understand the cyclicality of real estate but they deeply profoundly underestimate the magnitude and length of each cycle," he said. "Even at a top price, I could get a bargain. It was still a crisis price."

The start of recovery in Paris may well have green-lit the best building in Europe but this next period was much more to the industry than a return to bricks and mortar. By the time Coeur Défense was completed in 2001, the American investors had gained enough ground in Europe to seed an entirely new approach to property.

That early intervention in the Paris market had proved crucial. It enabled American capital to dominate the European investment landscape and begin to revolutionise the industry through its exported methods.

American investors were here to debunk the myth that property investment was only about intrinsic value, that worth stored in the bricks and mortar. Instead, they introduced the idea that financial value was based on future cashflows from buildings.

It began a movement that drew the industry increasingly away from the property at the heart of the deals. It even started to remove something of the soul about trading property; as a business there had always been a kinetic energy about property that never failed to get under the skin of those who came into contact with it. Now, for the American investors, it was, as one described "money coming out of boxes".

The Americans were untethered from the physical property, which enabled them to be so much more ambitious than traditional investors. The Europeans were always down on the street buying property; the Americans took to the skies to trade in the broader forces on which the industry spun. They studied macro-economic shifts that would open up major plays in countries and different strands of the property market. This was combined with that audacity to enter markets at historic lows, informed by repeating patterns of pricing, liquidity and the confidence brought about by cycles.

"Every time I've gone into a market that's coming out of a distressed cycle, the people in the market are always looking backward," said David Brush, who worked for Bankers Trust, later Deutsche Bank, at the time. "What happened last year is irrelevant to me. I'm trying to figure out what is going to happen over the next two years, which is why domestic players could never rationalise their own market. They get caught up in where they've come from."

It was this sort of thinking that gave American capital its entry point into Paris, and from the beginning, it was the reason for their very existence.

Back home, American investors had cut their teeth on distressed properties and portfolios of loans bought from the Resolution Trust Corporation (RTC), a fore-runner of the CDR that managed the loans in Paris. In the early 1990s, the RTC had mopped up the mess left by 767 defaulting US savings and loans associations after they overextended to business – real estate, in particular – leaving billions of dollars of buildings and non-performing loans in US government hands.

The US government needed to recoup at least some of its bailout money, but as the RTC was set up with a limited life, there was no time for a piecemeal approach. Instead, billion-dollar packages of buildings and loans were sold off at speed very cheaply.

This is where the American investment banks stepped in. Up until this point, they had not been significant buyers of property. The industry in the US was a private investor's game, individual plays only. But now that the RTC was offering up its wares on a wholesale basis, these sales moved outside the capabilities of the traditional property players and played instead into the hands of financial institutions.

Unfazed by the scale and deeply attracted by the pricing, the banks quickly established the methods to handle these transactions. They had the ability to reach around thousands of individual properties and loans and to price them collectively as portfolios. It also helped that they had the network and the swagger to convince clients to hand money over to them to buy them on their behalf.

The RTC provided such a mammoth opportunity to buy real estate at deep discounts that it accidentally spawned an entirely new industry. Before arriving in Paris, the American banks had swallowed up US$394 billion of domestic property and non-performing loans.[1] They paid, on average, 30 cents in the dollar, or one-third of the value. By 2002, the companies originating from this wave of investing were the largest private owners of real estate in the US.[2]

The unwinding of the RTC broke the investment banks into property, but it also opened their eyes to the longer-term potential of the asset class. They looked past the individual trading mindset of the traditional industry to see an untapped resource from which their aptitude for efficiency and scale could reap major profits. "I think there was a view that real estate was becoming an asset management business as much as it was an opportunistic business," said Russell Platt, who worked at Morgan Stanley and JE Roberts in this period. "It would have industrial scale and impact on the financial results of the companies that would be involved."

Property was starting to meet the ambitious standards of financial institutions, which expected their industry specialisms to harvest consistent earnings through fees, and build the stature of their companies by growing assets under management. It was also an area they could own beyond their existing client universe. "In real estate, firstly they didn't have the competition from their traditional private equity clientele because those folks had not yet gotten into real estate," said Platt. "Secondly, I think they came to believe on the results of their activities in the very late 1990s and part of the 2000s that they were uniquely competent and qualified."

## Capital with attitude

When it came to understanding this new entrant into the market, the reality was that not all sources of American money in Europe were the same. There were strong genetic links, like children of the same father, but with different temperaments and ambitions, they were far from homogenous.

The common link was their strategy to invest by taking on high levels of risk, an approach for which they found willing clients among their pension funds and insurance companies. "In the US, it was hard to find a pension fund that had less than 50% invested in equities. In Europe, it was hard to find a pension fund that had less than 50% invested in fixed income," said Marc Mogull, executive chairman of fund manager Benson Elliot. "One is risk tolerant, one is risk intolerant."

Back then, it was already embedded into the psyche of American institutional investors that risk was a spectrum to be played across. This was not like Europe, where their counterparts either invested in high-quality, low-risk property, or in no property at all.

US institutional capital had attitude. It had the flexibility and appetite to invest part of its money in higher-risk, higher-return opportunities while anchoring the majority of its investments into stabilised income-producing properties. In the mid-1990s, it was the adventurous end of its allocation that was in the hands of the American players as they came over to Europe. It was this inherent fluency with risk that meant that only American capital could have driven this change in Europe. "So, whilst we paint on a European canvas, we buy our paint elsewhere," said Mogull.

What divided the American investors was how much store they put in the potential of the property they invested in – or to see if from their side, which was the most financially lucrative.

At the furthest end of that spectrum was the investment banking model, which originated out of a world where there had been little or no direct experience investing in property. At best, they had advised clients on major property deals, but when it came to apply their skills to investing, deals for them had been solely financial engineering trades, or as one American fund manager put it: "They wouldn't know a piece of real estate if it came up and hit them in the head." This league of players were the familiar banking names: Morgan Stanley, Goldman Sachs and Lehman Brothers.

The investment banks structured their real estate business by setting up and investing through opportunity funds, the first of which were established to buy those portfolios coming out of the RTC. These funds pooled together money from US institutions, and often looped in their rich private clients, promising returns of 20%.

Soon after others with a stronger property heritage also adopted the opportunity-fund model. Firms such as Blackstone, Carlyle and Apollo set up opportunity funds that were soon active in Europe. These still had the same strong financial leanings but with more property expertise embedded into the strategy and the team. Finally, there were other players that employed the same methods but with their own money. When Bankers Trust and GE Capital invested, they did so from their own balance sheets, so-called principal lending. As a broad sweep, all of these were considered to be investing opportunistically, whether it was through funds or not. Later this evolved to be referred to as private equity investing.

But it was that purer financial style, the opportunity funds originating within the investment banks that really represented the essence of the change. The ideas – directly and indirectly – that spread across the industry in the next decade found their source here. This was the kid who always got the younger kids into trouble.

## A second phase of opportunity investing

With the sales of non-performing loans drying up, American capital was already on its second phase in Europe. From the outside, the transition may have appeared seamless but this was a whole new wave of people arriving to the continent.

By now, the non-performing loan specialists were long gone. They had shut their spreadsheets in Paris and were heading to overheating lending systems in Asia. "They left a sort of settlement, it was almost as if the Mongols came through, plundered and left some small genetic markers," said Platt.

In their place arrived bankers from the US who had built the opportunity funds business. "These were so-called veterans of the New York investment banking-driven model," said Platt. They supplemented investment banking teams that were already active on the ground doing property finance or advisory work.

The effect was to parachute purely financial people into an industry whose traditional focus on property was operational. "They were looking at the world completely differently," said Platt. They arrived with new techniques to analyse properties and to price and forecast deals. It was something they always described as new "technology", as though the industry had invented the BlackBerry, when really it just came down to processes.

The American investors fired up their own Excel spreadsheets to bring order and method to the large portfolios of buildings and loans for which they were now bidding. "Americans are very good at process. They invented the Ford Model T, and how to build the car that way. That's really in their DNA; how to take a task and break it up in sub-tasks and run a very efficient process," said François Trausch.

Squeezing buildings into this dry desktop analysis should have been awkward and unnatural, but without the same attachment to the bricks and mortar, there were no problems adjusting. Sure, they recognised valuable properties when they saw them but these were not priced on traditional metrics such as a square metre basis or the yield. Instead, the driving force for investment was cashflow. The security and longevity of the income stream was the central basis on which they assessed good property investments.

Taking this approach was breaking something cardinal about investing in property. As they were primarily motivated by the potential of the cashflow, American investors could afford to put much less emphasis on the quality of the bricks and mortar. It was a take that threw the traditional dynamics of the industry into disarray; did the quality of the buildings really matter if they produced a secure income?

For the American investors, this thinking was in step with their wider vision of making money from property. Driven by macroeconomics, their ideas drew them up above the fray of the buildings to capitalise on investment themes: capital dislocations, supply/demand imbalances and wholesale restructuring.

Then, pushed forward by the following winds of the economic recovery, this became a period when you could make money without knowing too much about the property at ground level. "What we were doing effectively was buying the market," said Mogull. "And, people would look at the US buyers and say 'boy, that was a lousy building they bought'. It's entirely possible no one even looked at the building because the issue was the Americans were buying the opportunity, not the asset."

But the Americans were not just in Europe with their money, they also came with advice. The big investment banks – Morgan Stanley, JP Morgan and Goldman Sachs among them – bolted on property teams to their established advisory businesses in Europe.

The idea was to offer the gamut of services that they did in other industries: mergers and acquisitions, initial public offerings or capital raising for listed companies and structured finance. Basically, to have much bigger ambitions for the property industry than it currently had for itself.

Their arrival on both fronts also anticipated another seismic shift in the region. Mocking Americans for assuming that Europe was one big place was about to end. This was the late 1990s and suddenly it wasn't naive to view Europe as one major property market, it was progressive, and the 11 countries just about to join the single currency were about to prove that.

The advent of the euro suited the mood of the Americans. It prompted a wholesale restructuring in Europe that connected with their thinking about the scale and financial potential of property.

A new layer of financial transparency across the eurozone was the perfect back-drop for price arbitrage on a grand scale. This was anticipated to be particularly fruitful with the expected convergence of low-cost economies such as Spain, Italy and Ireland, as well as those countries knocking on the European Union's door from Central Europe.

Meanwhile, the opening up of the financial borders was raising the game for major corporates, in particular national telecoms and utilities companies. Now, they had to sharpen their competitive edge on a European playing field.

In contrast, Europe's property players remained in national mindsets, shaking off the capital-starved lows of the early 1990s. A few had started to recognise the potential of the euro to property, but it was much more difficult to fathom revolution from the inside.

That delayed reaction from the home team was to make a big difference. As the European markets regrouped and recapitalised, that period of four or five years from 1997 provided American capital with a head start, a clear run to perfect their craft on European soil. Counter to everything held true about property, with a pan-European property market emerging and American investors blind to national boundaries, a window just opened up when you did not have to be local to succeed in property.

## Major shifts in property ownership

For the opportunity funds and similar US investors in Europe at this time, buying the market came down to two types of deals: corporate spin-offs and taking listed property companies private. Both these investing trends were out of reach of the average traditional European buyer, instead embodying everything emergent about American capital at the time: the fearlessness to tackle deals of such scale, financial engineering, and the clinical switch of emphasis from buying individual properties to a numbers game.

The corporate spin-offs were the first to show the true potential of the European real estate industry. In a single bounce, they pushed industry deals out of the hundreds of millions of euros bracket and into the billions. Still today, the EUR3 billion France Télécom deal in 2001 is one of the largest property deals France has ever seen. At the time, it was the second largest in Europe.

Right across Europe, American investors seized the opportunity to help major companies offload billion-euro property portfolios from their balance sheets in return for hard cash and an agreement to lease back the parts they still needed to occupy.

It was also the first instance of the American investment banking machine creating their own market, first on the advisory and then the investing side. "Yes, we sat down and thought about which companies were heavy in real estate and shouldn't be," said Dennis Lopez, who worked for JP Morgan at the time. "When we came to Europe we saw that many previously state-owned companies that had been spun off, owned all their real estate. Corporate ownership of real estate takes up a huge amount of capital and provides a very low return, especially relative to the much higher returns that were expected of a telecoms company." Later down the line, the US capital lined up to bid for the portfolios.

France Télécom was at the tail end of this raft of corporate spin-off deals that sprang up like forest fires across Europe in the four-year period from 1998 to 2001. In that time, around EUR37 billion[3] of corporate property changed ownership, mainly into American hands.

The corporate deals also cracked open the sides of the accepted European property universe. The spin-offs meant an influx of new buildings – good and bad – moving from isolated corporate ownership into the available stock that was being traded by the industry. That expansion itself was expected by many to seed future growth: the creation of giant property companies founded on the buildings that were coming from this great exchange.

Once again, the Americans looked back across the Atlantic for inspiration. Companies in the US owned around 18% of the properties from which they operated their business and leased the rest from landlords. In Europe, almost the opposite was true, with 70% in company hands.[4]

This side of the Atlantic, the state-owned telecom companies were an obvious target. Not only did they own most of their premises but they were in desperate need of cash to buy 3G mobile phone licences to remain competitive.

Philosophically, it made sense for the corporates to sell. This was capital locked up in property that could have been working harder for the core aims of the business, particularly as companies were struggling to raise cash in other ways. "You are already too levered so you can't issue more debt, but what's this EUR4 billion real estate portfolio on your books?" said Brush. "And, by the way, if it's on your books at 4 billion, and you've owned it forever, the market value is actually 7 billion. You can sell the real estate, raise capital, and book a nice big capital gain."

## Switching stations in France

Olivier Piani's position on the France Télécom deal was as much about the health of the company as it was the property. He was leading the GE Capital side as it put together a bid to buy EUR3 billion of property from the newly privatised French telecoms company.

Piani's rationale for the deal back in 2001 was to first consider the likelihood of the newly privatised buyer going bankrupt. If you understood France, Piani said, the probability of that was zero. "You have to explain this to the Americans [at GE] because they didn't always have the same view. For them AT&T can go bankrupt but France Télécom cannot go bankrupt. British Telecom cannot go bankrupt, no one would let it. They would restructure, find other solutions but it would not go bankrupt so this risk did not exist."

The second consideration was whether France Télécom would vacate all its buildings at the end of the agreed lease term, leaving the consortium as the owners of 475 empty buildings.

Again, Piani figured the chances were low. The fact that around half of these properties were technical properties such as switching stations complete with physical telephone lines feeding into them, was reassurance of that.

But Piani's thinking on this deal was an insight into how clinical the American investors' view was of the property at its heart. The risks that Piani faced were in the main financial, and at a property level, they were wholesale rather than focusing on individual buildings.

France Télécom's portfolio of technical properties actually made for an unnatural property portfolio. While centrally located in French cities and towns, the buildings were placed where they needed hubs to connect wires – locations driven by a functional need rather than a human one.

Transferred into specialist property hands, most of these properties would not have survived. They did not stack up as traditional property investments. Other than the odd artisan bar or café, these buildings were pretty much redundant beyond their specific telecoms use.

But with a focus on cashflow, American investors looked past this, or actually just didn't see it that way. To GE Capital and its fellow investors, France Télécom was the perfect portfolio.

They had the ability to see this as a financial transaction, with much less of a need to pour over future uses of buildings or ways to improve them to add value. There was no need. When the indefatigable France Télécom occupied them, it ascribed the future rental income a value way beyond the bricks and mortar.

The security of rental income from such a high investment grade tenant opened the door to cheaper financing, which meant they could maximise their returns on lower levels of equity. In this case, a syndicate of German banks which lent the consortium $2.2 billion, almost 75% of the cost of the $3 billion deal. GE Capital put in equity of $188 million,[5] with the remaining $612 million coming from the two partners, Goldman Sachs' Whitehall fund and French institutional investor Caisse des Dépôts. "We bought EUR3 billion of high-yielding assets, and once you have done this and already made a good deal, you maximise the leverage which makes it a fantastic deal," said Piani. Just three years later, the consortium refinanced the debt and GE Capital received back 50% of the equity it invested.

It changed the face of GE Capital, cementing the company's path to becoming one of Europe's largest real estate investors. Back then, its assets under management had been $3.7 billion,[6] a figure that grew to $30 billion globally.[7]

Corporate spin-offs were happening all across Europe but it was Italy that felt some of the greatest reverberations. Largely a closed shop to international property investment, the country's lack of transparency and liquidity (much of its property was held by Italian pension funds) had until this time deflected even the most pioneering early international investors. This made it even more ripe for the US opportunity funds; it could only be to their advantage to source deals under the cover of an opaque market.

Morgan Stanley made a first early entrance into the market in 1997 with a L200 billion (EUR103 million) deal to buy non-performing loans from Istituto Bancario San Paulo di Torino.[8] A couple of years later it paid EUR110 million for a portfolio with its new Italian partner Milano Centrale.[9] The amount was relatively meagre but it was already one of Italy's largest ever property deals.

But these early moves were soon bite-sized compared with the corporate spin-offs that took place over one month in 2000. First, Lehman Brothers teamed up with Italian property company Beni Stabili to buy 581 buildings from Telecom Italia. The partners took a 60% stake in IM.SER, a newly created listed company, which valued IM.SER at EUR2.9 billion, and allowed Telecom Italia to pay down its debts by EUR2.69 billion.[10]

Italy's new largest property deal was almost 25 times the size of the last Morgan Stanley deal, and instantly put Beni Stabili on the map as the biggest property company in Italy. It was also the second largest property deal that Europe had ever seen (later overtaken by France Télécom), and the third biggest globally.[11]

That same month, Goldman Sachs' Whitehall fund paid EUR620 million for a 90% stake in a similar listed spin-off company, Immobiliare Metanopoli, owned by Italian energy giant ENI, as well as a further EUR500 million on a second portfolio owned by the company.

The next year, the same changes were seen in Switzerland. In early 2001, Goldman Sachs' Whitehall fund became the first international investor to buy a Swiss property company since restrictions on foreign ownership were lifted two years previously.[12] Again, the entry deal was relatively small – 13 offices and retail properties in Geneva and Zurich – but it had cleared the path for Lehman Brothers to take the lion's share of Swisscom's CHF1.3 billion (EUR854 million) sell-off of 166 buildings in partnership with Swiss company PSP Swiss Property Group.[13]

These corporate spin-offs showcased the multifaceted approach of US investors on the larger deals, which became their signature style. "We saw that we could manage it better, we could get a good yield going in generally with long leases, the financing spreads were good because they were coming down so you had a nice leveraged cash on cash, and you had a lot of cases investment grade tenants you could borrow a lot against that, and then you had all the upside from you being able to increase cashflow just by managing these portfolios," said Brush. This included sales to local private buyers that would value the strong tenant highly. "It's a 15-year lease to ENEL in a building in your home town and the spreads could be as much as 150 [to] 200 points lower than the yield that we paid for it. So, you add all that up and those deals were very very profitable."

Running through this at all times was that ability for the opportunity buyers to price the risk of buildings of all qualities, divorcing themselves from the bricks and mortar thinking. "It was all underwritten asset by asset but in the telecoms deals the NPV [net present value] of the leases was by far the majority of the value. The predominant way to underwrite those deals was price the lease cashflows but be careful about the value of the conversions upon lease expiry," said Struan Robertson, who worked for Morgan Stanley at the time.

This was so counter to how the traditional investor would have viewed these properties. Without a solid plan to convert them into winners, they would not have bought low-quality buildings in the first place. But for the bankers, with a financial mindset, there was no expectation for these buildings after the end of the lease. They were prepared to let these buildings fail.

In the end, the underlying assets were properties that had to be managed, with the US investors often teaming up with or contracting this out to their local part-ners. For some companies, these deals were the springboard for them to grow into major pan-European players including French company Foncière des Régions, Italian companies Prelios (then, Pirelli Real Estate) and Beni Stabili (later part of Foncière des Régions), as well as strong domestic companies such as Corpus Sireo.

The timing was such that the early deals were very profitable, particularly playing all those different angles. For example, two years after buying 60% of the Telecom Italia portfolio, a consortium led by Lehman Brothers had disposed of 75 properties from a block of 190 mostly rented to Telecom Italia and sold the remaining 120 to Goldman Sachs' Whitehall fund for around EUR1 billion, a profit of EUR380 million.[14] Beni Stabili originally contributed EUR258 million for its 45% stake in the company,[15] which would make Lehman Brothers' contribution for its 15% stake around EUR89 million. Just the sale of these 120 buildings from the original 580-strong portfolio had returned Lehman EUR57 million in profits.

However, for the corporates it was more of a short-term fix as they under-estimated the lack of flexibility they were selling along with their properties, with many over-paying for 3G licences. "At the time, the strategic imperative to invest in their networks was critical. And their financing options were limited. In retro-spect the telecoms made long-term commitments to the sale and leasebacks that ended up having significant consequences as technology changed, interest rates fell and accounting treatment changed," said Robertson. "How many of them look back and wish they had found an alternative funding option? Many were under financial pressure because they had invested enormous sums for those 3G licences."

## Building and breaking the listed market

By all accounts, the American investment banks considered the public property market in Europe a mess; a repository for companies that had listed a long time ago and still happened to be hanging on in there. These were "lifestyle companies", "terribly inefficient", and certainly didn't fit in with the Americans' big-picture plans for the sector.

They were companies that had comfortable lives, judging themselves on the single metric of growth of net asset value (NAV), a rather blunt instrument that favoured a passive holding property rather than getting stuck in and actively managing it.

It was a shock for Floris Van Dijkum, who arrived in Europe to work for Morgan Stanley from New York where he was an analyst covering the US listed market. He expected Europe to quickly replicate the category-killer flotations for companies such as US property veteran Sam Zell's Equity Office Properties $4.8 billion initial public offering (IPO) in 1997,[16] or the creation through a merger of $10 billion retail giant Simon Property in 1998.[17]

Instead, he found that he missed even the basic information that was common for real estate investment trusts (REITs) in the US to report. "It was like pulling teeth, disclosure was not very good. Initially, we had a pretty big sector in the UK, and there were a lot of companies that frankly were not doing a good job in allocating capital wisely," said van Dijkum.

The saving grace across Europe was Sweden, which kick-started a wave of new entrants to the market. With a banking system weighed down by a similar post-crash pile of non-performing property loans, it had established a similar work-out company, Securum, to bring liquidity to the markets. Rather than wrap the non-performing loans up into portfolios and sell them off cheaply (the opportunity funds were poised and ready), Securum held onto them and got to work.

Five years later, in the spring of 1997, with a restructured portfolio and the markets turning in its favour, Securum floated 60% of Castellum, one of its companies, raising SKR1.6 billion to return to taxpayers' coffers.

It was bittersweet for Morgan Stanley, which advised Securum on the listing but lost out on its ideal plan to buy the portfolio for its opportunity fund. Securum had recalled its Scandinavian socially minded roots and decided it "didn't feel comfortable about giving away the store" to an opportunity fund.

The revival of the listed property market in Sweden was a much lighter shade of what the US investment banks had seen across the Atlantic but they recognised potential. In the US, the listed property markets were a major contributor to both the resurrection of property after the crash in the 1980s but also to its wider recognition as a financial asset class.

For major investors that still owned property after the savings and loans crisis in the US, they faced two possibilities: go bankrupt as they tried to refinance their portfolios in the US debt vacuum, or turn to the public markets to recapitalise. Most chose public life, listing themselves as REITs, a tax-efficient public company, and becoming part of a batch of flotations that saw the REIT market increase 12-fold from $10.8 billion in 1992 to $128 billion in 1997, the year Castellum floated.

This eagerness from US institutions and the public to buy REIT shares to get exposure to the property markets showed that the public market and its ability to be a source of capital for property was integral for the growth and health of the industry. "It was the beginning of the REITs in the US, and we need to be preparing for this in Europe," said Jon Zehner, who worked at JP Morgan at the time.

## Reinventing the listed sector

What began next was a systematic reinvention of Europe's listed sector, spurred on by the US vision. With the investment banks now in pursuit of the lucrative mergers and acquisitions (M&A) business around these IPOs, the lens through which the listed industry was being observed changed.

Bankers were walking into boardrooms of listed companies across the continent telling them that with the bankers' help, they could grow and be more dominant. "I will tell you that it got mixed reactions and we weren't trying to start with pan-European property companies," said Zehner. "There was still a lot of private capital around and you had to prove that the public capital was more competitive and that was a tough sell."

For an industry that was prone to cycles of overbuilding, crashes and resurrections, the transparency of the listed market was seen to provide a potential stabiliser. "I was a big believer that transparency was part of the issue and then to hear people like Zell talk about real estate being very capital intensive, and how access to capital is really important," said van Dijkum. "If you have a stock market listing then theoretically that should allow you to access the markets when you need capital and to grow your business going forward."

Some got it. In 1999, the European Public Real Estate Association (EPRA) brought together the leading names on the public side to improve transparency as a push to grow the EUR80 billion sector. The largest company, Land Securities at EUR6.6 billion, was dwarfed by oil giants such as Royal Dutch at EUR121 billion. With major fund managers favouring larger liquid stocks, the listed property sector would continue to be the preserve of specialist investors unless the landscape changed.

EPRA set about building an index to measure comparable performance across the sector and drew up reporting frameworks for consistency and to make the sector easier for investors to understand. But the largest impediment to growth remained the variety of national tax regimes with few countries having US REIT-like structures that underscored the rise of the listed market in the US.

Jeppe de Boer knew this all too well. Before joining Goldman Sachs, he was an analyst at ABN Amro covering the Dutch listed market. It was a useful grounding as, along with Belgium, the Netherlands was one of only two countries in Europe that had REIT-like listed companies.

The basic advantage for a REIT company was its tax efficiency. A REIT itself paid no tax and instead shareholders in the company paid tax on the dividends they received each year, with the REIT required to distribute a minimum level of its profits. These conditions were preferable to the double taxation that plagued the sector in other countries where the companies were liable to corporation tax and shareholders also paid tax on their dividends. The two layers of tax drastically reduced the appeal of investing in listed property companies, contributing to the generally stagnant markets.

With investment banks employing analysts such as van Dijkum and de Boer, companies now needed to fit into the European landscape rather than national borders. Most listed players were domestically focused, with only the Dutch companies having significant international investments. "I discovered that I could have a lot of value add just by crunching the numbers and trying to make the accounts of a French company comparable to a German company, comparable to a UK company," said de Boer.

This was old property laid bare and new property looking over its shoulder, but attention was not welcomed, including by the titans in the sector. "I can still show you a letter that John Ritblat [then CEO of British Land] sent to John Weinberg, the head of Goldman Sachs, basically asking him to have me removed for writing down such nonsense," said de Boer. "That tax couldn't possibly make a difference, and that everything was about NAV."

Across 1998 and 1999, Morgan Stanley and JP Morgan led the league tables by advising on a host of record-breaking IPOs. In the first year, the flotation of UNIM combined the corporate spin-off trend with the expansion of the listed markets. The largest listed company in Italy was launched with a L4,938 billion (EUR 255 million) share issue from Italian insurer INA.[18]

That same year, in addition to Castellum, Morgan Stanley took Sponda to the market to create Finland's largest listed property company.[19] "It seemed like we had a dominant market share in equity underwriting at the time," said van Dijkum. "So, [Morgan Stanley] could pick and choose which companies to target to take public."

Not to be outdone, banking rival JP Morgan advised Rodamco on splitting its global listed fund into four new listed companies to cover the UK, Asia, Continental Europe and the US in a EUR5.96 billion demerger.[20]

Across in Spain, Morgan Stanley also advised Inmobiliaria Colonial on its EUR442 million listing in Continental Europe's largest ever real estate IPO,[21] as well as an expanding Unibail on increasing its share capital by EUR247 million, to part finance a FFr6 billion property portfolio from Vivendi Group.[22]

But topping off 1999 for Morgan Stanley, was its highest-profile assignment to date, as it advised Canary Wharf, in Europe's largest ever real estate IPO, on the sale of new shares representing 25% of the enlarged company for more than £600 million, valuing the company at £2.6 billion.[23] The IPO, a testament to the growing appetite for dominant national players in the listed markets, was also an audacious comeback for Paul Reichmann. The Canadian property developer had crashed the Docklands development into receivership in 1992, only to buy it back for £800 million from the same banks that initially lent him the money. This time he was backed by a consortium of institutional investors and Saudi millionaire Prince al-Waleed bin Talal bin Abdulaziz al-Saud.[24]

The growth of the European listed sector was impressive but it was anything but stable. As quickly as the investment bank advisers built up the sector through new listings, the opportunity funds found reasons to take others private.

Rather than see its influence grow, the listed sector suddenly faced a period of instability that threw up doubts about the viability in Europe, let alone being able to replicate the success in the US.

Consolidation was expected and the UK, the leading market in the quoted sector, was already experiencing a shake-out of smaller and medium-sized companies. There had been more M&A activity in the UK market in 12 months to the summer of 2000 than there had been in the previous 12 years.

But when the news came that MEPC, the country's fourth largest property company, was to be bought and taken private in a £1.9 billion deal by GE Capital, it was a blow to the sector that few thought it would survive. By the end of 2000, 24 listed companies had either merged or completed takeovers in Europe worth EUR11 billion.[25]

It was always hard to be listed. On top of onerous reporting requirements and the scrutiny from analysts, the companies also lived at the whim of the wider stock market. Investor sentiment would either be with or against the property companies, regardless of their growth plans or current performance.

It was a problem that raised questions over whether the return from property company shares was more about wider market movements than the underlying property. The short answer was no, as long as you were investing over the long run. But success in listed markets was just like entering and exiting a windmill: you needed to get your timing right on both counts.

At this time, property companies were competing with other listed companies whose share prices were rocketing on the flimsiest of plans for new online ventures in the dotcom age. There was little interest in companies with physical assets paying rent.

Hopelessly out of vogue, the property companies were rightly disenchanted; on average, their shares were trading at prices that were 27% below the NAV.[26]

The discount was frustrating and, in investment terms, lacked common sense. The companies were being valued by the public markets at more than one-quarter less than the worth of its actual properties. In theory, it should have at least matched the NAV and a premium would be in order for good management. The industry was convinced the double taxation exacerbated this and continued to lobby the UK government for a UK REIT.

MEPC, for its part, courted its eventual buyer. It had already been hanging out with GE Capital for a couple of years. First, it sold them a £300 million UK portfolio[27] before teaming up to jointly own a second £200 million portfolio.[28]

By 2002, the company's management had had enough. It made every effort to meet the expectations of analysts: a new CEO, divesting its overseas assets, selling off its smaller buildings to focus on large office blocks, and selling or putting its retail assets into joint ventures. But still the price drifted down ever increasing the discount to NAV, standing at 440 pence (p) with NAV at 665p.[29]

That basic undervaluing of the listed companies was just too much of an easy arbitrage for the opportunity funds, as contrary as it was to the efforts of their advisory team colleagues to build up the listed sector as a vital part of the market. "It was frustrating to see it when we knew that over time the public markets had to be a bigger part of the answer, than they were even at the time," said Zehner.

GE Capital's bid for MEPC in partnership with UK institutional investors Hermes was agreed at 550p a share, a 17.3% discount to NAV, which narrowed to just 3% when taking into account the company's debt liabilities. Not only was it a good price for the property and the management, but MEPC's CEO said that as a private company, it would be able to run a higher level of debt and take more risks than was acceptable for a quoted firm. De-listing also moved management into the slipstream of US-style bonuses in the private market, with a controversial £65 million on the table if they outperformed.

Soon others followed as the US opportunity funds saw value in the companies trading at deep discounts to NAV. Alongside the corporate spin-offs, this formed a major part of their activities in this early period.

Deutsche Bank and JER partnered up to buy Allied London in a £138 million deal, while Deutsche Bank (which had by this point taken over Bankers Trust) and Grainger Property Trust bought residential developer Bradford Property Trust for £477 million.

On the continent, there was good news as well as bad: Rodamco Europe acquired Amvest while in Spain; Metrovacesa took over private company Gesinar. Meanwhile, opportunity funds were taking majority control of listed companies in operating partner investments in Spain and France. Colony Capital bought close to 100% of the shares of French company Lucia in 1999,[30] and Deutsche Bank took over 50% of Spanish company Filo's shares, upping its stake from 13%.[31]

The most controversial move, however, was UNIM. It lasted just nine months on the Italian exchange before Milano Centrale, the property arm of Italian industrial giant Pirelli, stepped in with a dramatic EUR2.3 billion bid in autumn 1999.

The deal was seen as a step back for the Italian property sector. UNIM's listing had heralded a positive move for transparency in the sector, particularly as many Italian peers had been trying to attract more international capital. UNIM was the public benchmark, already much happier to talk openly about its strategy and intentions relative to its new buyer Milano Centrale.

But what drew the most criticism was Milano Centrale's choice of investment banker: Morgan Stanley. The bank already knew UNIM well, having advised it on its spin out from INA. The investment bank's opportunity fund, MSREF, had also entered into a joint venture with the Milano Centrale earlier that year.[32]

When Milano Centrale won its bid for UNIM, it sold a 75% stake in the company's commercial portfolio to the Morgan Stanley Real Estate Fund (MSREF).[33] It was a move which raised the spectre of how well conflicts of interest were managed within investment banks when clients became targets of another part of the bank.

It also drew criticism from the bank's equities clients. "A couple of investors said 'look we don't like that'. Property investors said we are boycotting Morgan Stanley for six months. Investors were sensitive and it clearly impacted the trading volumes that we did, but bankers made a lot of money on mergers," said van Dijkum.

# 3

# FOR THE LOVE OF PROPERTY

It must have been love at first sight for those German investors, standing in front of the Lloyd's building in the City of London back in early 1996.

It's a property that wears its coat inside out, as if an old-fashioned counterfeit watch salesman had put his wares on show. Everything to run the building – the heating, plumbing, the lifts – are part of the protruding metalwork brazenly displayed on the outside of this silver wonderland.

For a nation of investors that found joy in bricks and mortar, and revelled in the dynamics of buying buildings, it must have been seductive for them to stand in Lime Street and glory at its inner workings. Of course, they wanted to buy it.

The Lloyd's building wasn't the first purchase by a German open-ended fund in London but at £180 million, it was the largest deal that a crushed UK market had seen in five years.[1] It was also an iconic architectural masterpiece. Its purchase by Despa sent London a clear message both about these buyers' financial firepower, and their intent to hone in on its finest properties.

The Lloyd's building was part of a run of big buys in London by the German open-ended funds. From this point on, they dominated their part of the European property scene. It resulted in an investing spree across Europe that automatically qualified them for a lead role in the industry's transformation.

Credit where credit is due, this set of 15 German open-ended fund managers were the first European investors to buy across the continent with scale and purpose. They reached far beyond their borders well before the single currency eased that path. These funds were the homegrown talent; Europe's very own masters of the property universe.

The Lloyd's building deal taking place in London in 1996 sounded just like the US investors: a bold buy in a broken market. But this is not a repeat of that story. The comparison is tempting because it's incredible to think that two such dichotomous types of investors walked the earth together.

Both were scouting around distressed markets in the early to mid-1990s, and their footprints led to the same locations. (To be honest, wherever the US investors went in Europe, the Germans were usually already there.) But there was one major difference: the returns. The Americans were seeking a 20% return, the Germans were looking for just 5%. They were the opposite ends of a poorly balanced property barbell.

The low returns should not make you underestimate the German funds. It took them less than a decade to go from purely domestic investors to owning EUR87 billion[2] of buildings in Europe and further afield.

The "why" was simple. It began to be impractical for the growing open-ended funds to invest only in Germany. It is always riskier to keep all your buildings in one basket.

It is the "how" that is much more fascinating, and the fact that they became so successful with a wholly exported – and often unorthodox – investment style. The way in which the Germans invested was so mired in their own domestic property culture that it should have struggled to survive outside its own habitat. But, unlike the Americans, the Germans' arrival in international markets was welcomed.

The funds arrived in London indifferent to the fact, or maybe collectively misunderstanding, that their domestic market differed radically from everyone else's. Yet they were not scoffed at by the locals or palmed off with leftover buildings. With the market in distress, they were lauded and courted, and offered the best buildings in town. Somehow these German funds thrived across Europe, and more than got away with acting like British tourists demanding a Sunday roast in Spain.

But what they did was smart, and it worked. They demonstrated that there could be successful investing on a mammoth scale at the lower end of the return spectrum. When it came to it, they succeeded because they loved buildings and only wanted to buy the best. No spreadsheets, no internal rates of return (IRRs), no calculated risks on mixed-bag portfolios, just pure, high-end, top-notch, for-the-love-of-it fine property.

If the American style of investing moved away from property, what happened with the German funds was the story of being too close to the bricks and mortar.

## Entering a devastated London

The majority of investors that swung the door open on the London market in the early 1990s immediately slammed it shut and gasped. It was a new kind of property hell out there.

Like Paris, London's property market had been devastated by its own strain of development fever, and the harbingers of this crash were in plain sight on the skyline. You just needed to count the number of construction cranes to know that there were too many new buildings coming onto the market. By 1990, the volume of new office space outpaced waning demand from companies looking to relocate or expand; rents and prices dropped, causing property returns to plummet by 27%.[3]

By 1992, the £1.6 billion[4] Canary Wharf estate had slipped through the fingers of property legend Paul Reichmann. Over in the City, property company Rosehaugh failed to pay back the £350 million it owed the banks, despite jointly owning Broadgate, the 32-acre City campus for the archetypal yuppie.[5]

The volatility of these times felled property legends, yet here were these German funds wandering around the city, seemingly indifferent to the recent storm.

It hadn't occurred to Despa and its compatriots to walk through the door on the London market to take in the whole city. Instead, they jostled for position with each other at the keyhole, giving them just enough of a gap to spot the Lloyd's building among the other beautiful prime properties up for sale in a devastated market.

DEGI, then the largest open-ended fund, kicked things off in 1991 when it bought Lloyds Chamber (no relation) for £72 million.[6] Other funds soon followed, and by 1994, the trend was unmistakable. Despa paid £104 million for 171 Victoria Street,[7] and close to £67.5 million for Hill House, in the City[8] and DIFA Spent £100 million plus on 6–8 Bishopsgate at Liverpool Street Station.[9] By the time Despa added 1 Lime Street, the official address of the Lloyd's building, to its portfolio for £180 million,[10] the Germans were starting to demonstrate their mettle as the original kings of prime.

All those deals were major for the time. Even today, two decades on, they would be considered more than respectable. The addresses may not be familiar but one thing these buildings all had in common was quality.

These deals reflected the German funds' hyper-focus on buildings. It sounds super-nerdy but actually it was a gift; an ability to put all their trust into the inherent value held in the bricks and mortar even in uncertain times.

In markets like London back then, their actions were like a child happily toddling along the edge of a cliff, but really it was a confidence to be admired. "That helped them to invest in markets like London because they really picked the quality assets, and they still do," said Barbara Knoflach, who worked for open-ended fund SEB Asset Management at the time.

What Knoflach identified was a purity to German fund managers' early behaviour. They could look through any market to the heart of the building, and that was the essence of buying prime.

With that sorted, they simply walked into London and bought top-class buildings at great prices when no one else was looking.

But the Germans' disregard for the volatility in the London market did not make the swings any less real. Certainly, their timing in the London markets was a misdirection that made them look more comfortable with risk than they actually were. "They generally like to invest when the market is less crowded. They are quite happy to be pioneers and look like they are taking relatively risky positions but they are not. They have a long term, income-focused investment horizon, and they are patient," said Robert Orr, who advised the German open-ended funds when he worked for property agent JLL.

The waiting implied a certain confidence. And, in most cases, waiting is exactly what they did. German fund managers were not natural sellers. They bought buildings and held onto them, using rents to feed returns to their investors. They certainly planned for a minimum ten-year hold, and with many buildings in Germany they were into their second or third decade of ownership.

That patience was important outside Germany. It would see them through the cycles, muting the noise of the volatility. Waiting drove a straight line through the waves of a market cycle. It suited them. They neither hankered for the highs nor wanted to end up paying heavily for the lows.

For this to work, the quality of the buildings was paramount. Performance centred on a small circle of property factors. Basically, a top-class building in an outstanding location pretty much guaranteed a solid, reliable tenant, which would pay the highest rent. The funds relied on very little other than the building for performance. They did use leverage but in very modest amounts, mainly to hedge any currency risk.

London was so sought after because it met all their criteria, and more. The UK capital offered quality buildings with top-rated tenants, on terms that the funds had never seen before. It was pretty standard for blue-chip companies to take 25-year leases. When Despa acquired the Lloyd's building, it was leased for 35 years at an annual rent of 11.6 million. The tenant, The Society of Lloyd's insurance market (in this case also the seller of the building), could only break that lease in the 25th or 30th year.[11]

## Investing the German way

Those basic dynamics were what informed the building blocks of German valuation. It was no coincidence that the Germans were focused on location, quality and a long-term hold when it came to their property. It was also no surprise that they were not natural cycle investors. Back home, they existed in an industry that had hardly any rhythm to it at all. Germany was an investment flatland.

Basically, no real market cycles. No volatility. No peaks, no troughs, no fear of overheating, no waiting for the ground to thaw. Every year property returns in Germany were about the same, hovering around the 6% mark. These consistent annual returns had a huge impact on the property landscape.

"You are not in a volatile market here in Germany," said Malcolm Morgan, who worked for the open-ended funds of DEGI and SEB. "If you are in a volatile market you have to think about where you are in the cycle. Here, in Germany, the only cycle that you had was basically the lettings side. On the capital side, it was generally pretty solid."

As Morgan explained, companies did come and go as tenants from offices and shops, so the demand for space did go up and down, but when it came to the capital value – the price of the building – very little changed year on year.

It's not clear whether the way properties were valued in Germany influenced the open-ended funds or the other way around. Some say the valuation system pampered the German open-ended funds, and reinforced the way in which they bought buildings. Equally, it could be the case that the valuation system pretty much dictated it.

German valuers had the same list of priorities as the funds: location, quality, lease length and rental income. To them, it was about a sustainable value, which when broken down revolves totally around the building's ability to produce a strong, stable income.

The valuers were more focused on the inside rather than the outside. They were more likely to be looking around the building, rather than out the window. This was because they were less interested in the outside market conditions that could affect the value of a building. For the valuers, it was as though, in isolation, a fine building with a long lease rose above all that is happening in the world around it.

If the economy fell dramatically in Germany, there was no quick re-rating of values. Instead valuers cushioned the impact by writing down values at a steady rate over a long period, often years. It was the same when the market improved. Values didn't naturally reflect the pace of optimism in the market, they just rose at an equally controlled rate.

For those who are not German, valuing this way is contentious. It's like valuation in denial. How can your building's value properly reflect the realities of the market, if you are not taking into account the wider economic changes? All other financial asset classes are being affected, so why not property? It was like the German economy and its property markets could fall off a cliff, but the value of any building would remain suspended in the air until the ground came back up to meet it. But this was just the way of life in German property, and the funds simply planted the right crops for the flatlands on which they harvested.

All this explains why the Germans came into London with such confidence. They saw quality buildings at prices so much lower than back home, and used a method that allowed them to see past the market conditions. "It's a little bit normal that the first thing you do is behave like you do at home, and they approached it with a German attitude. And then you start the learning process," said Knoflach.

It is also why the German funds held their properties for so long. To make money from the uplift from the market, you had to sell. Only then did you crystallise the gain from the rise in the price. The funds would never be opposed to a healthy bump in value from a rising market but no way were they going to depend on it.

Relatively, the rest of the industry lived and died by the market's volatility. The journey from peak of the cycle to its corresponding trough and back up again, well, that was the fun stuff, the market's energy. No fortunes are won or lost on a level playing field. But with the Germans planning to take this investing style right across Europe, maybe that was just the point.

## Unstoppable across Europe

As the German funds entered more and more new markets, they attempted to invest without fanfare. That wasn't easy to do when most of their deals broke new records, and they started to top investment league tables in many of European countries.

They were a permanent feature on the news pages of *EuroProperty*, but most of the stories were the dry bones of the deals. No congratulatory quote in the press release, no comments to put the deal in a bigger successful context.

The German funds were the dependable deal makers and soon became unstoppable, and not just in London. By 1996, in their two favourite cities alone, they had bought 29 buildings in London worth EUR1.7 billion and 21 in Amsterdam worth EUR360 million.[12] From a standing start five years previously, 11% of their capital was now invested in international markets.[13]

The push continued into 1997. CGI smashed a Dutch record when it paid EUR272 million for an office, retail and residential building in an Amsterdam suburb. It was the biggest deal the Dutch market had seen for 15 years.[14]

In the first three months of that year, they were already responsible for 50% of the £680 million invested in London by overseas buyers.[15] That's not bad for a breed of investors with their own uncompromising investment style that picked off the best from the skyline.

They were also broadening their horizons. Despa debuted in Spain while SEB (then BfG) proudly showed up in France, an entrance soon overshadowed by CGI's EUR200 million purchase of L'Etoile St Honoré, a *grand paquebot* or ocean liner of a building in central Paris.[16] Austria, their German-speaking neighbour, was soon put on the shopping list, while Brussels became catnip for the funds as they found it difficult to resist the gilt-edge covenants from European Union institutions looking to stay a long, long time in very large buildings.

But success meant the nature of the deals had to change. The open-ended funds were now investing in an improving market right across Europe. Prices were rising and yields were falling fast. In around the second wave of investing in 1995 onwards, when most funds arrived, they generally bought buildings for around a 9% yield. This was now slipping worryingly below 7%. The Lloyd's building was reportedly bought at a yield of 6.25%, causing a rival fund manager to tell *Estates Gazette* magazine that he was "flabbergasted".[17]

In theory, this downwards yield shift should have made London less attractive to the Germans. It was an indication that demand had driven up the price of the property and reduced the proportion of its income to the overall value of the building.

The fund managers also needed to think about subtracting fees and taxes on that income before it headed back to Germany. This stacked up easily when yields were at 8% or 9% (12% in 1991, those were happy days), but as they reached 7%, that was becoming more challenging.

However, the mathematics of it worried the Germans much less. They still looked to their own market as a reference point. Back home, they were used to paying very high prices, leaving more wriggle room in London. "In Germany, the market was basically the same for 30 years. So, for prime you paid 5% plus or minus 25 basis points and for the more 'B' locations, you paid 100 basis points more, no volatility," said Knoflach. "And, we were thinking, if we get into a market like London in a good location for 5.5%, then we are getting a good deal."

This made the Germans less in touch with the market pricing. Instead, they faced accusations of bidding up prices in London, and elsewhere. However, their early buys often paid off. CGI bought the two Atrium towers in Amsterdam's South Axis in 1997, at a 7.25% yield. Within a year, the yield had fallen to around 5%.

The Germans fired up the market but, as outsiders, they could afford to hold their own line. "It was exciting because they were moving the market," said John Slade, who worked as a property agent, advising German open-ended funds, including Despa when they bought the Lloyd's building. "They were giving liquidity to the market. Were *they* exciting? They were aggressive. Not as people but they were aggressive in terms of their buying policy."

Their moves into new markets was also usually strategic so to pay a premium for a high-quality building in London was never going to be a deal-breaker. The margin on the individual deal was less important than getting capital into markets with growth potential. "I remember a discussion with one of the German funds because we were giving what we thought was detailed analytical advice: if you buy at this price, it's a really good deal but if you get pushed too far, don't buy it," said Slade. What followed was the funds' reality: "'I take your advice on board but nevertheless I want to buy that building'. The German funds were buying a market exposure – the margin was not the issue. Entering into the London market was important, not the building's exact value."

## Moving into development

If the funds had a problem, it was supply. Doing £100 million plus deals offered them a privileged position to play in part of the market with so little competition, but it also meant few properties that were going to fit the requirements. The solution was a radical next step. Build their own.

If their major deals were vivid flashes on the property landscape, it was the developments they funded that provided them with a more permanent legacy. From 1995 onwards, they began to fund major office projects in London. By 1998, it was one of the activities for which they were best known.

It was a sign that the open-ended funds' strategy outside Germany was maturing as they gained experience in the market. Their development programmes also extended beyond London onto the Continent, as well as into other types of property such as shopping centres. And, once you passed it through a German fund filter, it made sense even from their more risk-averse perspective. "[Open-ended fund] investors didn't want more than 20% leasing risk on a completed building. They didn't mind on a development because that's normal," said Wenzel Hoberg, who worked for CGI at the time. "You build it, you have to lease it up but if you have the newest and the best kit in town it will lease up."

CGI epitomised the trend to build. Its first development was Milton & Shire House (now One Silk Street). It still sits in the north of the City where its 15 storeys of mottled brown granite butt up against one edge of the brutalist Barbican estate.

The £175 million redevelopment project picked up from the receivers for £50 million[18] was ambitious. Linklaters, the law firm, signed up in advance to be the tenant for the majority of the 42,500 sq m space, making it the largest pre-letting the City had ever seen.

CGI set the blueprint for development for its fellow funds. Under German law, the funds were heavily regulated as to how they operated, so typically the first fund manager into a new country or to stretch the boundaries of its activities such as development, paved the way by unpicking onerous legal and tax issues. This bred a collegiate and cordial set up where others were able to then follow, guided by the same documentation. The tight regulation was also another reason for funds' rather staid ways. Their rulebook was definitely more corset than belt or braces.

It made development an exercise in risk containment. For those early deals pre-lets were a must, or the prevailing lettings market needed to be rich enough that this was a mere formality. The buildings were extremely high-spec, but potential occupiers were blue-chip companies seeking new international headquarters. These were rare birds. "I think it was okay for the Germans to buy an empty development on a spec basis; it was not okay to buy a completed building that was vacant and hope for it to rent," Hoberg said.

Tough deals were also hammered out with development partners. Developers, by nature, are keen to agree their exit as soon as possible to move onto the next building. CGI's deal with UK developer Development Securities was based on an agreed yield so if the rents achieved were lower than anticipated, the price CGI paid for the building was also lowered. Meanwhile, Development Securities took on the construction risk, and signed up to share any development profits with CGI. If there was no profit to divide, it would pay back its development fee to CGI.

Its next project remained its most famous, and most British. Again, working with Development Securities, it smoked out the competition by bidding £50,123,456.78 for One Curzon Street in London's Mayfair, to demolish and rebuild the former MI5 spy headquarters. Again, records were broken. This 20,000 sq m office and residential project was the biggest office development ever seen in Mayfair.[19]

The contract for the sale of the building included a clause to stop the demolition works if they ran across a tunnel to Buckingham Palace, which is about half a mile away. If they found it, nobody said. Once the £200 million[20] project was complete, CGI and the developer shared in £45 million of development profits.

Milton & Shire House and One Curzon Street marked the beginning of a major development programme for CGI, which fast-tracked it to become a European rather than a German fund. By late 1997, the EUR7 billion fund[21] was more than half invested outside Germany.

With the rental market moving with them in most European cities, more funds turned to development. DIFA financed the £70 million Thames Court on the northern bank of the Thames in the City of London,[22] while across the Square Mile Despa signed up to 25 Chiswell Street for £140 million.[23]

On the Continent, that first buy for SEB in Paris was also a planned redevelopment,[24] and Despa headed to the sunshine to forward fund the 55,000 sq m shopping centre in US developer Hines' mammoth Diagonal Mar project in Barcelona. CGI topped this all off when in late 1999, it announced an EUR890 million development programme taking in Paris, Lisbon and Belfast.[25]

## The power of the German public

In the heat of so much spending by the German open-ended funds, it was easy to assume that the source of the money would always be there. However, in as early as 1997, this constant stream of cash looked to be in danger of drying up.

Net inflows into the funds for 1997 more than halved, dropping to DM6.4 billion (EUR3.3 billion) in 1997 from DM13.9 billion (EUR7.1 billion). Not only was the pace of the money coming in slowing but in the fourth quarter they saw a net DM260 million (EUR132 million) flow out.[26]

Overall, it had been an incredible growth period for the funds; the Müchener Institute estimated that more than 70% of the funds' cash had been collected in the five years between 1993 and 1997. (Bear in mind the funds had been going since 1959.) This surge had provided them with around EUR26.3 billion to invest[27] over that five-year period.

The figures around the German open-ended funds are astonishing enough in themselves, but even more so when you realise that the investors are not rich but just regular members of the German public. At the height of the funds' popularity, the man and woman on the German *straße* walked into banks and deposited an average of EUR286 million a week[28] into funds.

That sudden drop in money flowing in was a whole new experience. It opened up the possibility that there could be a run on the funds, causing net outflows. Not only was this the most significant reversal of money the sector had ever seen, it was happening during this new era of international investing, when the funds had reached an unprecedented scale.

For open-ended fund investors, property was always the default long-term investment. For some, that was a psychological assessment of what was considered successful in the decades after the war; those with land and property fared better than those without. For others, it reflected the Germans' innate conservatism; the tangibility of property made it the perfect safe haven investment for this prudent crowd.

German men or women would go into a bank on a Monday morning and buy shares in an open-ended fund. Then, on Tuesday, they could decide they'd had enough of property so head to the bank and redeem the shares to get their money back. That is what it means to be open-ended. With hundreds of thousands of individual unit holders in these funds, even with most investors' long-term outlook, that liquidity – the inflows and outflows of cash, whether to pay for a wedding or fund a retirement – was a daily reality.

Open-ended fund managers took these inflows and invested them into property, conversely the most illiquid of all investments. It defied property logic. You simply can't buy a building today and then sell it again tomorrow, and on a larger scale that was the shadow in which the fund managers invested.

"Basically, the daily inflows or monthly inflow statistics were like your lifeblood because it kind of indicated how big you wanted to go, how aggressive you wanted to go with your development programme or whether you would need to start thinking about sales, which came later," Hoberg said.

Fund managers confidently dismissed the 54% drop in inflows in 1997 as a blip, pointing to the funds' role in safe, long-term savings for the German people. Instead, they were more concerned about the backlog of cash, particularly as the right properties were getting increasingly difficult to find. Some effectively shut their doors on investors by telling the salespeople in banks – whose high commissions had been partly responsible for the growth in the first place – to restrict the sale of new shares because of the weight of money still to be invested.

By law, the funds had to keep a 5% liquidity cushion to satisfy investors' redemptions. In early 1998, the funds were awash with 20% to 30% in cash. It at least meant they were in a good position to satisfy those wanting to redeem their units, but in general the low returns on cash meant it only added to their problems.

At home, the property market was in recession. Despite the funds' expansion overseas, the majority of their properties were still in Germany. This was having a major impact on the funds' overall performance. For 1997, the funds posted total returns of between 3.6% and 5.7%.[29]

If there hadn't been an alternative, this may have mattered less. But Germans were turning their hands to the stock market. Seduced by the rise of the telecoms sector and the dotcom boom of the late 1990s, the usually cautious Germans wanted in on a market that was returning a heady 40%.

They had not found a new taste for risk. Instead, this was being encouraged by their government. They presumed they were in safe hands when buying shares in recently privatised state companies such as Deutsche Telekom.

These were new problems for the underperforming German open-ended funds, but despite the shocking year of poor returns and falling inflows, it did provide some clarity. It became obvious that funds with larger international investing programmes were easily among the best performers. For the 12 months to the end of February 1998, the funds of CGI, Despa, SEB and DIFA were among the front runners with returns of between 4.4% and 5.5% (the lowest was 3.5%).[30]

Meanwhile, it was the newly launched DespaEuropa, still relatively small, but with 80% guaranteed to be spent outside Germany, that returned 7.1%[31] – almost one-third more than its nearest rivals. German investors singled out this fund and any others with significant international programmes for new cash.

Investing in international markets was a real differentiator and one from which the fund managers could not afford to retreat. Drawing ahead of the international pack justified diversification, and confirmed the success of the open-ended fund model abroad.

Net inflows hit their lowest point in 1998 at EUR2.4 billion[32] but at the same time, the world's stock markets wavered in early 1998 in the run-up to the Russian devaluation of the rouble. With it came the first lesson for the German public in the volatility of investing in equities.

Investors flocked back to the funds and inflows almost tripled to EUR7.1 billion in 1999.[33] It was their second largest annual intake of cash ever, with the reverse volatility exposing a new danger. Now their size meant that the force of money flooding in was as dangerous to the funds as money pouring out.

Fund managers put cash into bonds to avoid breaching the 49% limit that could be held in cash, as they struggled to find suitable buildings to buy. Average fund performance was still hovering under 4% but it was clear from the money washing in that Germans preferred safety over performance.[34]

It was a watershed moment for the funds' managers, and they moved up a gear to meet the demands. In 1999, it was estimated that the funds had EUR5.1 billion to spend.[35]

This meant even more new markets and new types of property. It was about finding homes for the cash, but it was also about chasing yield. They needed to go further afield in less mature markets to find both the stock and the yields they needed.

This led them east, in an historic move. Early in 1999, Despa became the first to invest in Central Europe when it bought the Central Business Centre in Budapest for EUR25.56 million.[36] The deal was tiny alongside its others but it was truly significant. Before this, there simply was no institutional property market in Central Europe. It was a 10.5% yield, which elsewhere would have indicated a second-tier building, but double digits reflected the risk of investing in the best building in town in the small, emerging Hungarian market. The funds later set the same historic benchmarks in the Czech Republic and Poland.

In more established markets such as Paris, with its enduring appeal to the open-ended funds, capital was deployed in huge chunks. CGI paid EUR274 million for HRO International's Europlaza office project at La Défense.[37] Despa's pan-European fund paid more than EUR305 million to forward fund three projects for HRO International. All three developments were speculative and in the business districts of the suburbs of Paris.[38]

Then, early in 2000, Despa forward funded the 40,000 sq m redevelopment of the Crédit Lyonnais headquarters in Paris in a EUR328 million deal. Four years on from buying the Lloyd's building, Despa was funding the largest speculative development in the Paris market for the next three years.[39]

Then even more ambitious was Space Park in Bremen, Germany, a EUR316 million complex of shops, restaurants and bars – and a space rocket simulator. Despa forward funded it, the first time a German open-ended fund had invested in a major leisure project.

To get to this final phase of deals, they had stretched the boundaries of their earlier strict investment criteria. The ability to keep up their original investing style was being eroded by relentless inflows, as well as the incredible pace of change in the improving markets.

The volume of investing already raised the issue as to whether these later deals already displayed a lack of discipline by the fund managers in their eagerness to get the money away.

Of course, the other way to look at it was that they had adapted to the international markets, and it was working. Indeed, back in London they did something unexpected. They started to sell. The strength of sterling against the German mark, as well as the steep increase in the value of their buildings, was enough to make them reconsider their long-term holds. The proceeds helped shore-up returns back home when the fund performance was still flagging.

By 1999, funds owned £2.5 billion of property in the city, and now £1 billion of that was being sold or in play.[40] SEB began when it sold 70–88 Oxford Street, but it was DEGI that raised the game. It dispatched an eight-strong central London holding – around half of its UK portfolio – for £205 million to Gerard Ronson's property company Heron International.[41] It was thought that selling in London made DEGI a 30% profit from the currency shift alone.[42]

At the end of 2000, property agency JLL tallied up the investment volumes for the year. German investors, the majority of which were the open-ended funds, had spent EUR5.6 billion outside Germany, and EUR5.1 billion inside.[43] For the first time, the Germans spent more outside Germany than they did at home. In less than a decade, they had become the largest players in the European cross-border market.

# 4

# THE RETURN OF THE EUROPEANS

When the time came, the resurgence of the European property industry pivoted on the mundane matter of sheds. From the large characterless boxes at strategic motorway junctions to the back street multi-let industrial estates, this may have been a country's economic infrastructure but there was little that could be done to make the sector glamorous. That is, until late 1999 when Prologis changed all that.

The US company announced that it had raised a mammoth US$1.07 billion of equity for its first European fund. In total, 16 institutional investors stepped up to back the company's expansion plans for Europe. Six of them handed over more than $100 million each to the new fund.

But this fund wasn't in the style of the American opportunity funds that had blown into the markets just a few years before. In fact, there was no label for this type of fund yet. Instead, it was defined only by its characteristics: a 12% internal rate of return and a maximum 50% leverage ratio that gave it spending power of $2.7 billion.[1]

The money backing the Prologis fund was also not just American. Together, European investors accounted for 69% of the capital raised. Simple mathematics dictates that they were behind at least two of those $100 million big-ticket investments.

To raise chunks of money this size from European pension funds and insurance companies at this time was unprecedented. These were investors taking determined steps outside their domestic programme to back a campaign of international investing. After the last few years of observing their American counterparts from the sidelines, European investors were confidently stepping back onto the field.

It was not only the fact that they were back in the game, it was also the way in which they now chose to invest for the next cycle. After a long track record of direct investment – buying and owning their own buildings – institutional investors were going indirect, handing over their planholders' money to companies such as Prologis to be invested on their behalf.

This moved the investors one step away from the property. It was indirect investing, after all. The investments made by the institutions were into the Prologis fund, not directly into the property. These were financial products; property wrapped in an envelope, like mini individual companies each with their own corporate governance and reporting, and then run by fund managers for a fee.

This was it. By creating and investing in funds, the European property industry had just found its launch pad. By 2007, the Prologis fund was just one of 475 in Europe that together owned property worth EUR336 billion.[2]

Almost overnight, funds became the natural way – the only logical way – for institutions to invest in property internationally, if not at home. It also marked the birth of European property fund management as an industry. Today, fund managers are estimated to have around EUR1 trillion of assets under management in Europe across a range of private products including funds.[3]

Moving to indirect was a switch in style that had a profound effect on the growth and shape of the European property industry and one that completely rewrote the rules on the types of companies leading the business and the skillsets of those working in them.

Bundling up properties into funds was also the way to be taken more seriously by the rest of the financial industry. Stakes in funds were a more easily tradable product, and one that sat above the cumbersome layer of buying and selling of buildings. This fed into the industry's own desire to be a more liquid commodity, but also made property more appealing to a wider range of investors.

But there are always compromises. Whether it was noticed or not in the scramble to capitalise on this fast-evolving industry, going indirect slowly changed the very nature of property as an investment. Had anyone fully absorbed this at the time, they might have been more careful for what they wished.

## Awoken by the euro

The early entry by US and German investors into international investing in Europe had led them to dominate the growing cross-border market. In 1997, they were responsible for more than 60% of the $13 billion of international deals.[4] But this market share was now starting to wane as the early profiteering from bargain basement prices in capital-starved markets was coming to an end.

Already in 1997, European property had turned in its best performance for five years. It was "hardly stunning" with rental growth at 3.7%, but 1998 was already predicted to rise by a promising 7%.[5] That in itself was causing domestic buyers to stir, but in the end, it was the euro that finally brought them fully awake.

At the stroke of midnight on 1 January 1999, the arrival of the single currency liberated the investment landscape for European investors. The ability for buyers to look across different markets to assess and buy property unfettered by currency risk unburdened them both financially and psychologically.

With this smoothing out of the financial landscape, property yields were predicted to converge across the 11 countries, while the common interest rate would bring the cost of borrowing into line. The return that investors expected from property across the markets would become more uniform, a trait that was common in the other financial sectors.

That being said, property was not about to lose its character. Still firmly in place were the local lease structures, the complex legal and tax systems, and all those local quirks that made supply and demand factors in any one market unique.

This was a promising mix for European investors; one debilitating factor had evaporated but there was still much to chew on. What was a land of domestic country parts, was raising itself out of its recent crisis as a European whole. It was definitely good timing for the locals to make a comeback.

For the European institutions, the reasons to invest internationally were very different to those of their American peers. Going overseas for the Americans was tactical: high-risk, high-return investing funded from a final opportunistic slice of the capital they set aside for property. For the Europeans, it was strategic, a move to reallocate big pieces of their overall investing programme to non-domestic property. To say that it was not about higher returns is not quite right, but that certainly wasn't the intended starting point. For them, this was about the benefits of diversifying out of their domestic market, both to spread risk and to enjoy the fruits of different cycles.

For some European investors, the ability to break out of their national borders was moving from an opportunity to a necessity. The Netherlands, Sweden, Denmark and Finland were cornered in their own countries by the combination of a highly developed pension system and a small property universe.

As cash from growing numbers of planholders poured in, it stopped being prudent to make further investments into the limited pool of buildings in each of their countries. The teams managing property for pension funds were faced with a choice: drive up prices in their own backyard, or get wise to the opportunities in the international markets.

The Dutch were the first to tackle this problem, and it was their response that truly opened up the indirect market. They laid a path that enabled them to invest efficiently internationally which was also opened up to the hundreds of the smaller Dutch pension funds.

The Netherlands was the archetypal over-pensioned, under-propertied country. Already with ABP and PGGM, it boasted two of Europe's largest pension funds, both of which had been among those that had invested internationally since the 1970s. The Dutch purchases abroad had been sporadic; investing via flight plans, they were said to own the nearest buildings to the arrival gates for most KLM international flights.

But as the inefficiencies of managing buildings dotted around the world grew, ABP was the first to blink. By the end of 1994, it had retreated from its direct international investments, before rolling the same strategy out to its domestic holdings the next year.

It created Vesteda, an arm's-length property company, for its residential buildings, while the office and industrial portfolio were hived off into a new fund, KPN.

PGGM soon followed suit with its NLG7.5 billion portfolio. Abroad, PGGM took a 31% stake in US property company Cornerstone as part of a $1.1 billion deal to sell its US portfolio. On home ground, it merged the majority of its Dutch commercial portfolio with Aegon's to form a NLG5 billion jointly managed property fund. All of this relieved ABP and PGGM of the management burden for their investments (for a fee) while freeing up capital to make further indirect investments across new markets and sectors.

But it was a third institutional investor, Nationale Nederleden (NN), part of ING Group, which made a business out of its need to restructure. Faced with the same concentration problems, it not only outsourced the management of its existing investments but leveraged its property portfolio and in-house expertise to establish itself as a fund manager.

In setting up ING Real Estate (ING), the company joined the Dutch institutional investors expanding across Europe, but also manufactured the means to quickly become one of the industry's biggest players. In 1997, its assets under management totalled EUR2.7 billion. Ten years later, this had grown to EUR30 billion.[6]

The difference for NN was its place as part of the large financial company ING Group. NN had invested directly in property since the 1960s, building up a large portfolio of private and social housing stimulated by post-war tax breaks and subsidies. Meanwhile, the banking side of ING Group was a long-term lender to local shopkeepers and small business clients but had also become their landlord as it developed shopping centres in the expanding Dutch cities. Already, that development expertise was expanding overseas, building shopping centres as well as new offices abroad for the growing ING business.

In 1995, the company combined these different property activities to form ING Real Estate but as its investment activities continued to draw criticism, it made sense to harness these skills in a more commercial way. "Ultimately, they saw that real estate, for example, might be managing a couple of billion of property with 20 people, and equities might be doing ten times that with two people," said Pieter Hendrikse, who worked at ING in the period.

It didn't help that the culture of property people tended to rub the rest of the group the wrong way with their flash cars, brash suits, and desire for more money; "they have bigger mouths," as one Dutch fund manager remarked.

Four years later, the company embarked on Project Binnenstebuiten, or inside out, reversing out its property skills into a fund manager which would then service NN Group as its new client.

It partitioned NN's Dutch property holdings into three separate funds, instantly creating mammoth sector specialists, all of which still exist today. The NLG1.8 billion ING Dutch Office Fund, the NLG700 million ING Dutch Residential Fund, and finally merging NN Group's own shopping centre portfolio with an existing fund – Winkelfonds Nederland – to form the largest of the three, the NLG2.1 billion ING Dutch Retail Fund.[7]

Within three months, these funds were opened up for other investors (including ABP and PGGM) to buy shares in. This was the crucial step at which ING marked itself out as one of the earliest business models in property fund management.

ING sold NLG850 million in new shares[8] in the three funds to other Dutch investors, reducing NN's ownership. Taking on new investors moved ING from managing only captive money from within its company to investing third-party money for a fee. "Then, we are the fund manager and we get the extra investment money from the others," said Jan Doets, who worked at ING and led the transformation. "We invest that, and then we get the fees alongside those that [NN] pays." The money raised by NN was earmarked to expand its own investments internationally.

Considering the later frenzy of the funds market in Europe, this was a logical and level-headed beginning. Everything about ING's creation was sensible: modest returns of 8–9%, low debt and secure income, this was all very safe.

The opening up of the funds to other Dutch investors was almost collegiate, a solution for NN to redistribute its holdings as a means to diversify. This gave like-minded, mostly smaller, investors keys to an unexpected kingdom: access to top-end properties, often of a size and quality that they could never afford to invest in on their own.

NN retained a large stake in these funds, and subsequent fund launches, providing a natural alignment between the performance from ING for its in-house client and its other investors. "It was a solution with diversification, and with an operator on the ground who knew what they were talking about as they had done a lot of direct investment in the past," said Hendrikse.

With a portfolio of buildings on hand to seed its early funds, and guaranteed investors for new launches, ING was on the fast track to becoming a market leader. Its early moves made the company instrumental in rallying the Dutch investor community to take the first steps into indirect investing, and setting them on the path to being the most sought-after source of capital for funds in Europe.

As ING established its ground, others across Europe were also leading to the setting up of some of today's larger fund management houses. In the UK, Henderson Global Investors opened up an office in Frankfurt on the back of bringing German clients to London, and followed that up with France and Italy, the foundation of its Continental European network. "Our initial aim was to bring German and then European capital to the UK. Investors were attracted by the transparency, availability of research and the safety of the market," said Patrick Bushnell, who worked for Henderson Global Investors at the time.

In the early 2000s, its role as a major fund manager was cemented when the UK insurance companies found themselves under pressure, in part due to the dotcom crash. The closed Life funds were shut to new investments, and, as they naturally reduced in scale as policies matured, it was decided to switch significantly from shares to bonds, and also to reduce property holdings. Many turned to straight property sale, but Henderson offered its insurance company clients an alternative path. "We suggested that they put some of the properties into Jersey Property Unit Trusts. We set up a number of sector specialist funds in which they could retain their participation and sell over time. They also achieved increased diversification and we attracted other institutional capital and expanded the portfolios," said Bushnell.

Like ING REIM, this resulted in the setting up of cornerstone funds including the Central London Office Fund, the UK Shopping Centre Fund and the UK Retail Warehouse Fund, all of which still exist today.

Meanwhile, in France, CDC IXIS Immo, the first iteration of what was eventually to become fund manager AEW Europe, was formed when the French institutional investor Caisse des Dépôts (CDC) externalised the management of its property investments in 1994. It managed CDC's portfolio and then in 1999 launched its first fund, IXIS Logistis, which specialised in the new generation of high-class big-box warehousing now demanded by sophisticated distribution giants. This was soon followed by Fondis, which invested in high street retail across France.

## Building category killers

As a fund manager, Prologis quickly established itself as a category killer in the logistics sector, first in the US and then Europe. Before it even raised its fund, it had moved Pac-Man-like across Europe, hunting down the best logistics parks. In the summer of 1998, it bought Kingspark, a renowned UK logistics specialist, for $157 million.[9] Four months later it swallowed whole Garanor, France's top distribution park, for $317 million.

With these and subsequent purchases in Spain, Prologis already owned the finest distribution parks in Europe (the fund would be seeded with $350 million of these assets), and had laid down a network of local expertise to retain its pre-eminence in each market.

It was this determinedly top-tier approach that separated Prologis out as the leader in its field, and being best in class appealed to investors. If they were going to place their money in the hands of fund managers, it needed to be for an investment strategy that surpassed their own abilities, and gave them access to properties which they were unable to get hold of themselves. As an American company, Prologis did this with particular gusto, which was why in a heartbeat, it raised $1.07 billion.

This specialist focus was also driving European fund managers. ING's Dutch funds and the UK funds from Henderson might have been in the fund managers' home markets but for investors they represented access to specialist properties, and the specialised management that came along with it. These were large properties like dominant shopping centres, which already had a scarcity value back then, and are an endangered species today.

Those funds starting from scratch also kept to the specialist track. Logistics was in vogue with AEW's first fund, and other launches tended to narrow in by geography and sector for pioneering areas: French high street, Polish offices, anything that was difficult for investors to access or in which they had little or no experience. "That's how the European fund model first emerged," said Rob Wilkinson, CEO of AEW Europe. "Initially it focused on new or specialist asset classes such as logistics or retail. Many investors were capable of investing in offices themselves so didn't need a fund to access this type of asset."

The shift to fund management also prompted changes in the type of companies and people needed to manage these funds. The majority of property work in the industry was transactional, offering a deal-by-deal service for those looking to buy and sell property.

Property lagged other financial sectors when it came to how it serviced its clients. In the UK, for example, property somehow sidestepped the financial Big Bang of the 1980s, which had deregulated the financial markets and kick-started a professionalised investment management industry in bonds and equities. "I can clearly recall going on the road to talk to local authority pension funds in Wales and to ask if they would be interested in us providing an investment management service for their property portfolio for a fee. And they would say 'No, we use a stockbroker (who in fact charged a commission but not explicitly) and they do it for nothing'," said Richard Plummer, who worked for PRICOA Property Investment Management.

The move to set up fund managers was about advice but it was also about discretion over the capital they were investing for their clients. "From a bottom line perspective, it was more fee generative for the investment managers and also it made the whole industry operate more effectively because you could move quickly," said Gabi Stein, who worked for Henderson and LaSalle Investment Management during this period. "You had professionals making the decisions rather than going back to pension funds where the person in charge of decision making was not necessarily a real estate professional."

The potential for long-running repeatable fees soon filtered through to the property agents, which were overly dependent on transaction fees. Now, under the influence of new American owners, many started to turn their attention to the fund management model.

With the cross-border investment market lighting up, the list of multinational companies looking for space globally was expanding, making the network of European offices that UK agents had built up since the 1970s hugely valuable.

Within two years, the four top UK agents were bought by US companies beginning the rise of the mega-agency as they also picked off strong national agencies such as Bourdais in France. But it also produced a less studied change for the European property fund management market.

The closing of the largest merger to date hit the markets in 1999 as the industry was at the MIPIM property fair, its annual gathering in Cannes.

Across the Riviera city, one-third of the 200-year-old Jones Lang Wootton brand was completely wiped out as the new Jones Lang LaSalle (JLL) flag was raised on flagpoles along the beachfront. Chicago-based LaSalle Partners had just paid $435 million[10] for the company, and from it created not one but two businesses. Jones Lang LaSalle for the agency work, but also LaSalle Investment Management, a new fund management business, bringing together everything that was discretionary investment advice rather than transactions.

The same occurred on a smaller scale within what today is CBRE, as it began its confusing but voracious string of mergers to become today's global leader. First, in

1997 a split of ownership – and opinions – saw the UK firm Richard Ellis divided by two US takeover bids. Insignia bought the Richard Ellis UK business and CB Commercial began its 20-year acquisition trail in Europe by buying up Richard Ellis' Continental European operations. (It eventually bought Insignia as well.) Soon after, the CB Commercial side combined its investment management businesses to form CBRE Investors, bringing together the separate account activities in the UK and US as well as its nascent funds business stateside. These restructurings created new delineated fund manager businesses but also new clients for the agency side, with LaSalle and CBRE Investors maintaining an arm's-length distance from their transactional cousins.

## The other side of Prologis

Colin Campbell saw the turning point for European property during a deal that he was working on for an American investor. It was a shopping centre in a city northeast of Milan that he had failed to persuade his American client to buy on a 9% yield, just a couple of notches down from the 9.75% that was their final offer. "Within a year, CGI [the German open-ended fund] had come and bought the property at about 6%. The difference was that we were about to buy it for EUR95 million and CGI bought it for EUR200m," he said.

To Campbell it was clear that the market had transitioned into a different phase, and to one where a greater understanding of the workings of the local market was needed to uncover value. The simple phase of cheap property was over and this was now an insider's game. "My response was that I needed a European organisation and secondly a European platform. Our model from day one was having Spaniards in Spain and Italians in Italy," he said.

In 2000, he joined Pradera, which had been set up by Paul Whight, a serial property entrepreneur who had successfully rolled out the retail warehouse format in the UK with his company Granchester. Now, Whight wanted to do the same in Continental Europe.

To them, Prologis had cracked it with its ground-breaking mega European fund. "The first thing I said was who owns the Prologis document, because if they raised a billion for warehousing, we ought to be able to raise something for retail," said Campbell.

The document was the private placement memorandum, the guiding rules for the fund. In the same way the German open-ended funds eased into countries once the legal hurdles were overcome, it was also natural for new funds to avoid reinvention if they saw others successfully raising money. "I think the discussion was five minutes about what we were going to do because we'd seen that Prologis was very successful and we wanted to do the same thing."

What was different about Prologis' fund was that it was a Luxembourg Fonds Commun de Placement (FCP), the first time that a fund was domiciled in Luxembourg or that such a company structure had been used for property. Even on a simple marketing level, it added to the European flavour of Pradera's offer, with Campbell keen to play down "being two Brits abroad".

More critically, it allowed for the free flow of capital for investment from a wide range of countries, a mix that was usually tripped up by the different tax treatments for international investors. "The whole point is that there was a precedent, and it worked for American money and Middle Eastern money, why do anything else?" said Campbell.

Domiciling the fund in Luxembourg was about to open the door for Pradera and a whole new generation of property funds to move up the risk curve and attract more than one nationality of capital.

As early as 1992, Plummer and Robert Gilchrist had set up the TransEuropean 1 fund to bring US capital to Europe. It was one of the earliest European funds, and also the first time a limited partnership structure was used for property, despite the structure dating back to 1907. "Until we fell upon this, we couldn't figure out how to achieve similar tax advantages for foreign institutional investors that property unit trusts enjoyed in this country for UK pension funds while, of necessity, co-investing alongside each other," said Plummer.

Other than the mixing of several nationalities, the limited partnership had similar advantages to the FCP: tax transparency and a flexibility to be an envelope for a group of investors, known as limited partners (LPs), to then be managed by a general partner (GP). "It taxed limited partners, not the product. That was the key, the golden key," said Plummer.

For Campbell, the Prologis fund had other new concepts embedded into its workings. The fund was inflected with ideas borrowed from its American opportunity fund cousins. While earlier models like those from ING charged solely a management fee, the Prologis fund introduced the carry.

This was a performance fee that saw the fund manager share in some of the profits once a certain return hurdle was reached. It was a new way to incentivise fund managers, giving them an enhanced status in the relationship with their clients. "You suddenly thought here's a model where for the first time the adding of value is properly rewarded rather than us just being fee-earning professionals," said Campbell.

The carry – or promote, as it was also known – was already common in the US. Plummer had already been tripped up when he'd been questioned on it in one of his first meetings with a potential client in the US. "We didn't really understand the concept of a 'promote' as, at the time, such a notion was highly unusual in Europe. And, that first meeting was an absolute baptism of fire. It was our first visit on property business as we were just not acquainted with the terminology or fee characteristics," he said.

But prior to the Prologis fund, Europe wasn't yet ready for this added complication. It was already hard enough to navigate Europe's multiple currencies, myriad tax regimes, and fragmented markets without then adopting complicated fee structures. "We were pioneering in an enormous institutional market in which we had some contacts but we had no familiarity with performance-related fees. We had nothing to base it on," said Plummer. "It was also difficult to make progress with performance fees given the prospective effects of exchange rates (this was a pre-euro era). Perhaps, more importantly, there was no generally accepted European benchmark. In the end, we decided it was much more important to launch TransEuropean 1 successfully on the basis of an all-inclusive fund management fee of 1% than not at all. Mind you, the [US institutions] couldn't believe that we didn't charge transaction fees or performance fees."

So far, Prologis had demonstrated how to unite different investors in one pan-European fund and also shown fund managers their worth as advisers, but ultimately it was the broader message of the fund that was to have the greatest impact. This was essentially a EUR1.2 billion boast that returns in the mid-teens were there for the picking in Europe, and that as a financial product, debt was instrumental in achieving that.

What imitator funds picked up on was that there was a fertile middle ground in the funds' landscape that was yet to be mined, a place in the geological cross-section which married up the elements of the higher risk opportunity funds' financial flair with a strategy more rooted in bricks and mortar.

Unfortunately, in a fast-accelerating market, it was also seen by fund managers and investors as a green light to be more ambitious about returns for all funds and to form an unhealthy reliance on debt to get them there. The fact that this next evolutionary step collided with an easy period for secure debt was only further proof that this was how these new financial products worked.

For some, the starting point for this new middle ground came from opportunity fund investing. In Paris, LaSalle's Francilienne fund was a pioneering early starter, already operating below the radar of the large US investment bank opportunity funds in the late 1990s.

Francilienne was opportunistic, an idea formed off the back of LaSalle's hard graft working out the first portfolio of non-performing loans in the city. "It was an evolution in our strategy from being a distressed debt player to one of buying direct assets at distressed pricing where we could create value by, for example, leasing up buildings with vacancy," said Jeff Jacobson, Global CEO of LaSalle Investment Management.

With 75% leverage and hitting that sweet spot in the cycle in Paris, Francilienne is said to have returned its four investors more than double its 20% target return in just two years.

Francilienne's successor at LaSalle was not another round of opportunistic investing but a fund that reflected that there was value to be found by employing leverage but leaning more on the properties for returns. "We knew it was a different style of investing for traditional institutional investors in the market but it was a natural evolution of things that were driven by this incredible opportunity in France. The challenge at the time was figuring out how do you find the money for that?" said Jacobson.

LaSalle's answer was Euro 5, named for the five Western European countries in which it would invest, and one of the first in Europe to give the style a name, "value-added investment".[11]

Placing more focus on the properties that underpinned the value-add funds should have been a good evolution for the industry. It naturally understood the concept of value add. Property is one of the few financial instruments where you can influence its value through hard work. Put property in the hands of professionals, and growing value through refurbishments, releasing and redevelopment is what they do best. This was the side of the equation that the industry understood.

It was this angle that Ric Lewis wanted to exploit when he arrived in London from the US to set up Curzon Global Partners, a joint venture between US firm AEW and property consultants DTZ. Rather than call it value add, he preferred a second, middle-ground classification, core plus. "We were basically taking things that have one thing wrong – a leasing risk, minor renovation, a credit problem – fixing that problem, making it more institutional, fixing the income stream and then putting it in the hands of someone who is a less active manager who needs duration and yield," he said.

It also fed into the notion that Europe needed its own version of business higher up the risk spectrum to attract local capital. "The mindset was different," said Simon Martin, who worked for Curzon. "To build a business we needed to raise capital locally and what was going to appeal to European institutions? At that point put an opportunity fund in front of European investors and they said 'no way'. We bridged European investors out through towards that opportunity." Curzon's second fund would have an opportunistic strategy.

What the industry had not yet upgraded to was a fluent enough understanding of the other side of the equation for the value-add funds. "[You've] got to understand both sides of the real estate balance sheet," said Wilkinson. "You have to be able to understand the asset side, that's the real estate itself, but also the liability side – meaning the capital structure – as both of these drive the returns in the end. If you had the ability to understand both of those, it is also true that you could go quite far quite quickly because there was a lack of that skillset."

Around the same time, Francilienne paved the way for a second wave of European funds that stayed with the opportunistic theme at the very end of the 1990s. It was still mainly US investors, but ones interested in piggy-backing on the more local approach. "They wanted on-the-ground managers rather than global partners with European allocations," said Noel Manns, one of the founding partners of Europa Capital.

Europa had wanted to raise money for private equity-style investing but its usual port of call – the listed market – was off limits as the market was out of favour. The team tried to tap into the UK institutions but these did not yet have the nuanced understanding that Europa found among their US counterparts. "[In the US] you walked down the corridor into the alternatives office, down to the room that said real estate and in the corner of real estate there would be a guy that said private funds," said Manns. "And, they knew what you were talking about: value add/opportunistic, pan-European real estate. 'Yes, we do that, come and tell me your story'."

Europa raised EUR600 million of equity for its EUR1 billion fund with ambitions of a 25% IRR, with at least 30% of the money to be invested in Central Europe.[12]

Europa were not the only ones. Marc Mogull joined Doughty Hanson, a UK fund manager, which raised $700 million of equity for a $2 billion fund, of which half came from US institutional investors. Again, the emphasis was on local professionals on the ground. And three partners who had started the LaSalle business in Europe – Van Stults, Aref Lahham and Bruce Bossom – spun out to form Orion Capital Managers, which raised $400 million of equity,[13] while Keith Breslauer left Lehman Brothers to set up Patron Capital.

In Europe, this move from conservative to higher-risk investing was the first conscious intersection of property and finance, leading the funds away from their benign beginnings of focusing on prime property and income. "Those first five years of funds were very much about the real estate guys doing real estate deals but they just happened to be niche markets or more specialised sectors," said Joe Valente, who worked for DTZ during this period. "It wasn't pushing the envelope of the capital stack, it wasn't using more debt, it was simply going into self-storage facilities or investing in retail warehousing for the first time."

Both the Euro 5 and Pradera funds launched in the European market within months of each other. Euro 5 raised $282 million of equity in 2000 to invest in office, retail and industrial properties in areas with "above-average growth" in France, Italy, Portugal, Spain and Germany.[14]

Pradera raised EUR600 million for shopping centres or retail warehousing across the continent, focusing mainly on fully let properties or by forward funding projects under development.

The two used higher levels of leverage than had been seen in funds in Europe before. Euro 5 could borrow up to 65%, effectively giving it $800 million of firepower and meaning it would commit an average of 35% of equity to each deal. Pradera was more conservative at 50% debt, giving it a maximum potential portfolio size of EUR1.2 billion.

These debt levels were considered mild at best. Campbell said Pradera was marketed as "core", an emerging label for the lowest risk and return type of fund on the market. "You would probably call it value add now. It was limited to 50% debt and it had a fairly tight range of core-ish European countries," said Campbell.

Fund managers were entering a world where greater returns seemed plausible and the accompanying risks perfectly manageable. "The value-added business in the US – which was not that much older than in Europe – was evolving and growing very fast. All of us who were in the US were seeing that these funds could get bigger, that they were attracting more investors, that there was a real need for this capital," said Jacobson. "Consequently, during that time the early participants in the market pushed along their ambitions faster than you might otherwise."

# 5

# A RISING RISK IN EUROPE

Coeur Défense finally took her place on the Paris skyline in the autumn of 2001. The market had easily absorbed the grande dame's record-breaking size, opening with just 6% unlet.[1]

In the end, her arrival was unexpectedly quiet; a September opening party was cancelled as the terrible events of 9/11 unfolded in the US. Its landmark status was secured but the timing was not right for Unibail to introduce its new twin towers to the world.

The terrorist attacks had exacerbated a volatile time for the markets. The new century had started off well with European property markets back with a bang. At 27%, returns were the best for a decade, and average rental growth had risen 22% in one year compared with the average 2% seen right through the 1990s.[2]

Much of the leasing activity was driven by telecoms and technology with the proliferation of mobile phones as well as the optimistic first wave of businesses off the back of the expansion of the internet. Young, untested companies, funded by excessive speculation in tech stocks, took up swaths of office space, driving vacancy down to the lowest levels since 1988.[3]

Then, that bubble of tech stocks quickly burst, stemming the main source of new office take-up, particularly in Stockholm and Madrid. It was a mis-step by the "new economy" but one that filtered right through to the old.

Take-up plummeted 40% in Paris and 27% in Frankfurt, while rents fell in Paris, London, Frankfurt, Madrid and Stockholm.[4] By the summer, rental growth was frozen, as global growth slowed, on the back of the tech stock dive and nervousness following the 9/11 terrorist attacks in the US.

This was the first shock of the new cycle but things had been so good in 2000, that most in the market merely dismissed the unexpected dip as tempering unsustainable levels of rental growth, and taking the froth off the market.

However, within that period, there was a more fundamental shift. A time when the market first split, the moment when the occupancy and investment markets started to pull away from each other, reducing their interdependencies for the rest of the cycle.

This seems like an early point in the cycle to mark the path towards the crash, but the fact that the industry rebounded from this short-lived downturn was the first indication. Certainly, there were some serious impacts locally for the markets most affected, but the net effect of the dotcom fallout for European property was positive.

As an asset class, property's ratings started to soar. Along with the regret of the money backing the overhyped tech companies had come another reminder of the volatility of investing in stocks. Soon, property's reliability was back in favour. For the next year, the occupancy markets remained sluggish but all eyes were on the investment activity which continued to be buoyant. In 2002, there was a 20% fall in leasing activity but investment volumes of EUR67.5 billion fell just 1%.[5]

Already, that weight of money looking for a more secure home was holding prices firm rather than prospects for rents and take-up. From this point on, the market would start to gain momentum, driving up prices and encouraging investors to take on more risk.

## European funds take off

As the news spread of successful fundraising, others saw in funds a winning formula that was not difficult to replicate. "Success begets success, but it also begets imitators and variations," said Jeff Jacobson. "If you look at people in Europe, the idea of a fund manager was 'Wow – I could raise big pools of capital that I have discretion over, that have reasonable fees and I can get a share of the profits'."

It was a lightly regulated industry, often only defined by the rules in the jurisdictions such as Jersey, Guernsey and Luxembourg where funds were domiciled. "We used to joke, two men, a dog and a basket of capital, that's all you needed," said Ric Lewis.

The funds industry did not need to be told twice that their product could absorb the growing levels of capital in the market. To satisfy growing demand, funds quickly became bigger and more ambitious. In mid-2001, UK insurer Standard Life launched a EUR500 million fund, seeded with a large chunk of its own properties on the Continent.

Standard Life had been an exception to its UK investor peers; an unheralded hero of early investing on the Continent, shrewdly buying in Paris and Brussels from the mid-1990s onwards. Now, it was taking the opportunity to wrap all those buildings up into a fund and sell it on to other investors.

The same year, ioGroup launched its EUR500 million industrial fund, LaSalle launched its successor to Euro 5, the EUR500 million Euro Growth II,[6] while French insurer AXA teamed up with IVG, a listed German company, to set up an EUR800 million fund to invest in the Paris market.[7]

New European opportunity funds were also filling out the layer below the financially engineered strategies of the investment banks. The Carlyle Group raised a EUR1.8 billion fund,[8] while PRICOA, part of the US Prudential, raised US$600 million of equity to invest in operating platforms such as self-storage and parking.[9]

In 2003, the billion-euro core funds arrived. Tishman Speyer was first with its European Strategic Office Fund, targeting European core on a scale that had never been seen before. These types of funds were usually smaller or more domestically focused. It raised EUR600 million, which rose to EUR1.2 billion[10] with 50% leverage, the levels of debt also being seen in value-add funds at this time.

Tishman Speyer had pipped Morgan Stanley, which then closed its EUR1 billion core fund, a wildly different – and tamer – animal than its opportunity funds. The core fund signalled the investment bank's intentions to be a full-spectrum player in the property sector. While its opportunity funds shot for 20%, this fund investing in Europe's main cities anticipated returns of 8% on an annual basis with the equity of EUR1 billion leveraged by 40%.[11]

The funds also ventured further east. AEW (then branded as CDC Ixis Immo) teamed up with Ergo to spend up to EUR500 million in Central Europe.[12] Heitman, now almost a veteran having made its first buy in Central Europe in 1996, was now pushing into Eastern Europe and the Baltics with a EUR650 million fund.[13]

The low barriers to entry saw an explosion of funds during this period. By the end of 2002, it was thought there could have been as many as 500 funds in the market with a total of EUR250 billion of assets under management.[14] Fund managers emerged from all parts of the industry: institutional investors, property agents, banks, as well as the two men and a dog varieties. However, the most prolific channel was the developers, which were already building property right across Europe.

In this flourishing market, the move for developers to tap the institutional funds market was obvious. In quieter times, it was possible – and prudent – to grow deal by deal, but as the markets accelerated, the only route to success was to build up a pool of discretionary capital. "When the markets start to move more quickly, having discretionary money at your fingertips means you can move quickly and get deals done quickly," said Gabi Stein, who was now working for Tishman Speyer.

It also allowed these long-standing private companies such as Tishman Speyer and Hines to avoid tapping the listed markets for cash and losing their independence.[15]

The trend was even more relevant for retail. Pan-European retail developers had spent their time from the late 1990s onwards building shopping centres of all formats across Europe. They were the forward party for the rest of the industry, already pushing east and south into undershopped markets in Southern and Central Europe.

From France and the Netherlands, food retailers such as Carrefour and Ahold were beating the fastest path east in pursuit of new food markets with their hypermarket-anchored shopping centres. From the UK came Hammerson, Pillar and Freeport, as well as Grosvenor, which part-owned Portuguese shopping centre master Sonae Inmobiliaria and had plans to expand out to Spain, Italy, Austria and Greece.[16] Then from the Netherlands came the development arm of ING, as well as Rodamco, Amstelland and MDC (which later merged).

By the early 2000s, there were 3,700 shopping centres in Europe over 5,000 sq m, as well as the emerging formats of retail warehouse parks or factory outlet centres that were springing up like molehills around them.[17]

As shiny new centres rolled off the conveyor belt, they were a good match for what investors wanted from their international investments. There was a low supply of the same product in Western Europe, and by heading east and south, they could get the same quality with higher yields (from 6.5% in Spain, to 9% to 10% in Poland).

Their size and specialist nature made funds a good fit for shopping centres. Developers seeded funds with completed developments, giving them a partial exit for their investments. They then retained a stake in the fund and took a management fee from the fund's investors. For those developers without any shopping centres in hand, raising a fund gave them a new source of capital, from those willing to back their expertise.

Sonae launched its first Sierra fund in 2003 seeded with 18 shopping centres valued at EUR2.4 billion,[18] while Amstelland-MDC used a fund to break into Central Europe, gaining instant management exposure when hypermarket owner Ahold seeded a fund with around 12 of its shopping centres worth EUR400 million.[19]

## Bringing the investors on board

As the funds industry moved through the early years of the new millennium, all the arguments were in place for investors to adopt indirect investing on a major scale.

Within the institutional investors, the spotlight was turned on property, and why it was the only part of its overall portfolio that remained domestic. "I think pension funds looked at their equity and fixed income portfolios and recognised that many years before they had gone international," said Neil Turner, who worked for Schroder Real Estate Investment Management at the time.

For fund managers, with funds, it was suddenly easier to sell different ideas to investors. They could test the waters of investing in other countries helped by the fund manager, moving some of their property eggs out of one basket. "The UK real estate market may be more transparent than a lot of markets but with the best will in the world, it's 6 to 7% of the global real estate market. Why would you miss out on 93% of the universe?" said Turner. "That's a pretty powerful argument, and a lot of investors bought it for very understandable reasons."

Modern portfolio theory had washed through the other asset classes and was now seeping into property. As they looked to maximise returns from property within comfortable risk parameters, the diversification they needed to achieve this could be gained from funds and the scope for international investing.

Collectively the industry saw the power that the funds had to galvanise capital from investors in a way that no other property investment strategy had been able to do before. Fund managers were no longer reliant for money on just the long-standing investors; armed with a new product and a better story, they were able to open the doors to new investors.

Every fund manager employed capital raisers, to market their own funds to prospective investors, and together these formed an army that was promoting the wider merits of investing in property indirectly through these vehicles. "There was a structural shift of pension funds coming into property. [Funds] brought in a lot of capital that hadn't thought about property before," said Monica O'Neill, who worked for Curzon at this time.

The setting up of the European Association for Investors in Non-Listed Real Estate Vehicles (INREV), an industry association particularly focused on the funds, added to the sense of a joined-up part of the market. There was work to be done to even measure the basics of this new industry, such as getting its arms around the size of the industry. "We were surprised how many funds there were out there, as the INREV team brought transparency to the market and compiled the database," said Pieter Hendrikse, the co-founder of INREV. By the end of 2003, just as a starting point, firms had volunteered information on the size and characteristics of 196 funds representing EUR154 billion of assets under management.[20]

INREV quickly followed up with a series of initiatives to map the market, and tools to give its industry the transparency and credibility to sell its wares to more and more investors: analysis of fees, volumes of capital being raised, as well as guidelines for fund managers on good corporate governance and reporting. It also provided a home for these new ideas and new cast of characters. Five-star pan-European events became the place to be for the fund manager, the capital raiser, the fund accountant and the much sought-after institutional investor.

The offices of investors became a carousel of fund managers spinning through, pitching their latest funds. The richest vein of investors was the Dutch, so capital raisers were passing each on travelators in Schiphol Airport, but equally bumping into each other in the airport's business lounges across Europe, and then increasingly the US, the Middle East and Asia.

For some investors, the single currency helped with a spirit of adventure after investing in the same finite markets. "It was interesting and exciting and provided diversity for them in the same currency because overlaying all this was the arrival of the euro and that just completely changed the risk profile for all these people," said Colin Campbell.

For the reluctant, the growing number of their competitors taking the leap started to show through in performance. Investment Property Databank was revolutionary in first providing property performance information, while INREV began an index solely focused on fund performance. "[Many investors] were unwilling to go out of their comfort zone because they personally were not going to be rewarded to take the risk. It's only when their own core real estate wasn't [performing] enough that they had to go out and get higher returns," said O'Neill.

Going international meant new structures, taxes, leasing schedules, and less job security than staying at home. To get comfortable, investors would apply what they knew about property rather than the product. "Experienced real estate investors knew how to underwrite a deal so they spent a lot of time going through the seed deals and wanting to understand the deal and whether we seemed like we were good real estate investors," said one fund manager. "They were never asking questions about the portfolio or risk management, bringing it up to the higher level."

There was also a tendency to let the bigger investors conduct the due diligence with smaller ones dropping into the slipstream, joining up once major players such as ABP and PGGM were signed up. "To a large extent, I think they used to trust and rely that the fund manager to whom they were giving money had the expertise that basically entitled them to make decisions with very little restraints on them," said Stein. "Some of them would literally just phone and would not ask any questions and within a month send you a subscription agreement. That is unheard of now."

Those selecting value-added funds knew this was a new higher-risk strategy and therefore understood the need for higher returns. Danish pension fund ATP committed EUR50 million to an early value-added fund as its debut indirect investment. "We went into this to buy assets, create value and sell, and that was a different model for how pension funds normally operate," said Michael Nielsen, chief executive of Danish investor ATP's real estate division.

However, more worryingly, investors fixated on requiring a premium on a core return when investing outside their own country. This was a counter-intuitive move that would leave the fund managers taking more risks in international markets to get those returns. "I don't quite think people had worked out quite what the risk premium should be of people going abroad. It wasn't very sophisticated," said Noel Manns.

More fundamentally, it showed that when it came to one of the main reasons, a chance to spread their risk by investing in different countries, diversification was a concept that investors did not wholeheartedly believe in.

It should have been one of the funds' biggest selling points, but fund managers soon recognised that to raise funds, there was a commercial reality. "I remember at the time thinking, I know I have to go along with that argument because I'm never going to win mandates if I don't," said Turner. "It had to have a 10 in it, a double-digit return. Whatever it was, it just kept going up. And, perhaps that is one of the reasons why we ended up with the products we did."

## Blurring of the lines for funds

Soon the pages of *EuroProperty* were writ large with stories that showcased the metrics of the age: not how much fund managers had invested in property but how many millions they had raised. The industry was on double speed as everything converged: the positive outlook for returns, property being recognised by more institutions and the rise in the number of funds. "I think there was a sense of make sure you don't get left behind. This was the future. We're all doing it, this is going to be brilliant," said Turner.

A part of the industry that had started out building funds for investors to share risk and take their first steps to invest overseas together was now detaching itself from those original conservative aims. It too was getting caught up in the momentum of the market.

Funds got bigger, broader while return targets increased and debt levels began to rise. "The demand was so great that people did then start to launch vehicles that to some extent you wonder how necessary they were; pan-European office funds and more generic style vehicles," said Rob Wilkinson.

Leverage for all funds, including core, was creeping up. Those funds at the conservative end now had at least 50% loan to value (LTV), the level at which the first early value-added funds had pitched themselves in the market. Value add with its mid-teen returns was getting closer to 60%.

Aiming to benefit from higher yields, fund managers also pushed further east and south as their usual turf in the established markets of Western Europe started to feel the weight of competition and capital.

The lure of a ready supply of newly developed shopping centres at higher yields opened up more opportunities in Spain and Portugal. With the office market still in the doldrums, newcomers including international funds from Germany, the UK and the Netherlands instead went on the hunt for dominant shopping centres and retail warehousing developments.

Meanwhile, returns in Central and Eastern Europe were predicted to outstrip Western Europe until at least 2007,[21] as the much newer markets matured apace. The region had the added attraction of seeing Poland, Hungary and the Czech Republic among ten new countries joining the European Union that year.

In early 2004, Apollo Real Estate Advisors showed the potential of the region when it sealed the region's biggest deal, paying $750 million to Metro in a sale and leaseback of 28 shopping centres in Poland.[22] The yield was around 10.5%, already making this a more challenging market for opportunity funds (depending on debt levels), but suited the value and core funds coming in behind them.

The funds started to follow this geographic expansion, as established core players such as ING moved into new locations, and up the risk spectrum. Its EUR600 million ING Property Fund Central Europe had a return target of 13.5% and 60% leverage. Its sister fund, the ING Retail Property Partnership Southern Europe, was bigger with spending power of EUR1 billion, of which 65% was debt, as it looked to achieve returns of between 13% and 15%.[23]

Hines was also moving into value add with a fund that raised EUR400 million of equity with similar returns of 13% and leveraging of up to 60%. US fund manager Heitman, which had led the charge in Central Europe with its early opportunity funds for the region, was now moving back down the risk spectrum, raising EUR100 million of equity to target Western European logistics with up to 75% of debt, in a fund that it was calling core plus.

All this activity changed the general direction of the funds, moving them away from their conservative roots, taking them much further up the risk spectrum. "Everyone was converging on a mean," said Campbell. "And the mean was a bit less financial than the Americans had been and a bit more than the Europeans had been."

# 6

# A NEW RELATIONSHIP WITH DEBT

Fast forward for a moment to 2010, two years after Lehman Brothers' fall marked the start of the global financial crisis. Hans Vrensen, then global head of research for property agent DTZ, had been asked to speak at the IPF/IPD Conference, a mainstay of the annual UK property calendar. The event was high profile, and content-wise, more cerebral than its many counterparts; packs of researchers, strategists and fund managers decamping to Brighton on the England's south coast to ponder the impact of the J-curve or debate observations from someone's earnest regression analysis.

Vrensen was presenting on the outlook across the European property markets, a miserable enough mandate in the wake of the crisis, but one that he relished. "I was pretty excited about it. But in hindsight one could see it was a bit of a trap," he said. "Of course, they saw me as the CMBS [commercial mortgage-backed securities] guy, and they were going to give me grief for all the things we had done."

With the presentation over, Vrensen took the hit in the Q&A as questions were fired at him about the bad alchemy of CMBS, the part of the lending market in which he had worked prior to the crisis. In that moment, he was a lightning rod for the audience's anger at the destructive nature of debt that was still wiping out the values of buildings right across Europe.

Vrensen stood his ground. "It takes two to tango," he told them. "You took the debt, and you did it because you were bidding each other aggressively in the market. Prices were going up and the only way you could make your equity returns work was to get lots of debt at cheap rates, and CMBS was part of that."

That audience's misplaced frustrations towards Vrensen were representative of a shell-shocked European industry. Embarrassed by their own disfigured businesses, everyone knew they had been too easy fooled by finance. Debt had not just been cheap and available, it had also been complex, hidden beneath layers of detail and obfuscation that all but a few were afraid to admit they didn't understand.

CMBS was the focus of that complexity, but the overuse of debt was as much about traditional bank financing as it was about this new entrant to the market. The amount of debt in the European property market in 2007, was estimated at EUR1.74 trillion.[1] It remains a rough figure, and that was part of the problem. No one was accounting for the systemic rise in lending in this latest cycle.

The industry has had a long relationship with debt but what evolved from the mid-1990s into the 2000s was about much more than the availability or even the quantum of debt. It was the same central character, but definitely not the same story.

Everything about how the industry used debt transformed over this period. Until the mid-1990s, debt was more long term, with short-term bridging finance largely a tool for developers needing financial capacity to ensure their projects got off the ground, or fuel – when available and cheap enough – for high-risk speculative property traders on the sidelines of the industry.

Now, in just over a decade it was mainstream. Regardless of risk appetite or return requirements, debt was endemic, working its way into the lives and liabilities right across the whole spectrum of property players who deal by deal became increasingly seduced by its power to enhance returns.

Bankers, pushed on by debt's newfound prevalence, plotted to stretch lending's limitations both in use and complexity. Through financial innovation, they seized upon new ways to lend more cleverly that ultimately – in the pursuit of market share and profit – just enabled them to lend much more, and badly.

The contribution of debt to this cycle stopped being about whether the tap was on or off and instead became multifaceted with new factors that would influence the cost and availability of debt. It was about banks working and competing internationally, a stronger drive for returns on equity and innovation that allowed banks to more easily parcel up and sell off risk into the capital markets.

As all this happened, there was not time to consider how the greater penetration of debt would affect the dynamics of the investors. With innovation and expansion came an intensified property cycle where debt – not equity – led pricing, obscuring the fundamentals of the real estate beneath it. The question was how did debt go from shadowing the equity to overshadowing the whole industry?

## Setting up the European networks

James Raynor was just 26 when he was dispatched to Paris to set up the Royal Bank of Scotland's (RBS) first lending division outside the UK, in 1998. Paris was chosen as it had come out top in the bank's research, as much for its market size as the fact that French banks were still in bad shape from the crisis. "They weren't capable of lending a great deal so we went over there really on a semi-speculative basis to see if we could get a market position – me and that was it," Raynor said. "There was no grand strategic plan, but we saw French banks had reasonably sized teams but no balance sheets, whereas we didn't have any team but a balance sheet."

There, RBS started on its path to becoming one of the most aggressive and profligate lenders in the run-up to the crisis. With second UK bank HBOS, it was responsible for 50% of commercial bank lending in the UK prior to the crisis,[2] and dominated the European lending scene along with two major banks from Germany.

Raynor's five years with the bank in Paris hit the intersection of broken French banks and the first wave of US players entering the city. At first, the French banks fronted deals, desperate to be seen to be back in the market despite having a limited balance sheet. RBS stood behind them, sharing the load by taking on a slice of the loan in a syndication. Raynor's first deal in Paris was a syndicated piece of a loan originated by French bank Société Générale as part of Prologis' US$317 million purchase of the Garanor distribution park.

This strategy gave RBS market position, and market intelligence. Raynor noticed that anyone doing deals of note was Anglo-Saxon: Goldman Sachs' Whitehall fund, Morgan Stanley Real Estate Fund (MSREF), GE Capital, Apollo, AIG and Blackstone. Soon he was building up his loan book by dealing with them directly. "Really what they wanted was to transact with someone in English, with the ability to turn around stuff pretty fast and for it to be simple," said Raynor.

The first wave of lending was conservative; most of RBS' business in France was based on loans 170 basis points over Euribor, the average interest rate at which banks were lending, on 60–70% LTV so "reasonably safe and easy money given it was also quite early in the cycle". But soon deals grew, as did the numbers of borrowers. "What became seen as a big deal changed. If it was EUR200 million in the late 1990s, that was a massive deal, but over time EUR200 million wasn't seen as big anymore. To start with most of our loans were EUR30–40 million, these just grew and grew," said Raynor.

By the time US investors settled in as clients, the first wave of European non-listed funds rolled in. "Suddenly you had ING as a fund manager come into the market. You certainly saw a lot more core players from about 2001 to 2002 and they made up a big chunk of the market," said Raynor, of this early profitable business. "The cost to income ratio of my business was seriously low and it was very profitable. We had running costs of 5–10% of income. Then, you had more hands out wanting to borrow and naturally all of the lenders got bigger and bigger as there was more and more business."

RBS' biggest rivals on the European banking scene were the German mortgage banks. Their first international outpost had been the UK, for some as early as the late 1980s. This had provided them with much-needed solace from a chronically over-banked market back home. The German lending scene was hugely competitive and as yet untouched by consolidation. Even today, in a much slimmed-down industry, there are more bank branches in Germany than petrol stations.

German mortgage banks were active lenders with a competitive edge. They could raise money for their loans by issuing *pfandbriefe*, highly secure bonds that gave them access to low-cost funding. Once regulations permitted them to lend on foreign assets, they took this competitive lending to a wider universe of much more profitable markets. "Margins were substantially higher outside of Germany than inside, so it was a combination of increasing diversification of their business and less concentration risk on purely local markets," said Jürgen Fenk, who worked at Hypo Real Estate at this time.

By 2001, around half of the mortgage banks' EUR12.5 billion of new loans was outside Germany, with international property lending accounting for one-fifth of their overall loan book.[3] The UK, France, the Netherlands and Spain kept them at their busiest, but they were open for business for local buyers and the growing number of cross-border players from Finland to Spain, Ireland to Greece.

Soon, greater returns on equity were prioritised as a growth target for the mortgage banks rather than traditional loan book size. This prompted a wave of consolidation from which emerged the German giants of lending that would dominate in Europe through the remainder of the cycle.

Eurohypo, the mortgage bank arm of Deutsche Bank, announced in late 2001 a three-way merger that swept up the mortgage bank businesses of Commerzbank and Dresdner Bank. That gave Eurohypo combined assets of EUR236 billion,[4] of which around EUR90 billion was in commercial property.

Meanwhile, HypoVereinsbank (HVB) created HVB Real Estate from three of its group's mortgage banks as well as the property lending business of its parent company. Soon after, a second shake-up saw HVB rebrand to Hypo Real Estate, a three-pronged company that listed on the Frankfurt Stock Exchange: Dublin-based Hypo Real Estate International for non-domestic lending activities as well as two German-focused mortgage banks.

In a third restructuring, Depfa group split into two, decoupling the property lending business from its public sector lending business (the only other activity financed by issuing *pfandbriefe*). The property lending side became Aareal Bank, a bank with an around EUR27 billion loan book across Europe, including its early lending in Central Europe.[5]

For all three, the restructuring re-emphasised a focus away from Germany and onto international markets, and a chance to service US capital and the growing flows of cross-border money.

Eurohypo was active in 13 different countries, writing 50% of new business internationally, and Aareal operated in 19 countries. Hypo had significant loans books in the UK and the Netherlands, and was building up critical mass in Spain, Italy and Sweden.[6]

The banks soon recognised that to meet profit targets and remain relevant was now much more about greater risk-taking and dynamism. Hypo, Eurohypo and Aareal were upfront about seeking out size and complexity to earn quick-turnaround fees rather than just expanding the loan book. Bankers, like fund managers, were reinventing themselves, moving from lenders of standard senior debt to specialist experts and advisers to international players on large and complex transactions.

Servicing more demanding borrowers was only a small part of their business but it soon drew banks' eyeline higher up the risk spectrum, making them much more comfortable with greater levels of lending risk. As businesses, they were in part responding to borrowers' appetite for more debt and at greater risk levels, but, in the name of a better return on their own equity, they were driving it as well.

All the banks strove to be more nimble with their balance sheet, recycling equity for further loans and fees by syndicating the larger loans they had taken on. While Hypo ruled out equity investing, it took on "equity-type" risk, or subordinated loans, the higher-risk loans that sat behind the senior debt in the pecking order if the loan defaulted.

Other services were bolted on to supplement and complement standard lending, which was still the bulk of the business. Eurohypo's new investment banking arm was supplying consultancy and structuring work for M&As, and providing bridge finance. Aareal Bank began a fund management arm, to invest third-party capital from institutional investors.

Peter Denton was number three to arrive in a revived property lending team at Deutsche Bank. The acquisition by Deutsche Bank in 1999 of Bankers Trust, the Navy SEALs of the higher-return investing world, brought the German investment bank up close and personal to the cutting edge of this opportunity funds world, and its banking needs. "There was a feeling that the market was getting better but was incredibly immature and more cerebral solutions were going to be needed to deal with the things going on," said Denton.

It required European banks to adapt, and to be in line for this business, they needed to be less domestically focused. "You had this sudden explosion of private equity money [mainly US capital] and banks started to look internationally rather than a cross-referral system. Before if I was in London, I would refer a deal to my French or German colleagues," said Denton. Now, he was sitting in London, part of one of the first teams that specialised in providing the finance to the growing number of large and complicated deals taking place across Europe. This was mezzanine lending, funding "take privates" from the listed market as well as lending on corporate sales and leasebacks. Deutsche Bank, historically a traditional lending bank, provided financing for many of the headline deals such as Lehman's CHF1.3 billion (EUR854 million) Swisscom deal in Switzerland, as well as its SKR4.9 billion acquisition of listed Swedish property company Tornet in the largest deal of 2003.[7]

These were what Denton calls "the golden years of lending" – that period from 1999 to 2003 when banking was truly innovating in a changing property industry. "Yes, and we made very strong profits, not at the expense of the market, we were just doing a really good job. These were lending activities that were needed and it was fun, exciting and stimulating," he said.

## Origins of the CMBS market

At first, European banks were adamant. They were not interested in securitisation. The idea of packaging up their loans and selling them to investors as a series of bonds just did not appeal. This was just not how the industry liked to lend its money. It preferred relationship banking, the cosy long-standing ties with borrowers that produced repeated business, connections they were reluctant to jeopardise. There was no way they were going to swap that personal style to offer loans that would be taken off their books, shredded up into bonds and sold into the fixed income market as CMBS.

Besides, the banking industry didn't need it. Germany's banks alone could have provided the markets with enough debt, and as the leading lending force in Europe, its depth, breadth and tight margins did not leave much room for new lending techniques.

None of that was going to stop the US investment bankers trying. In the US, CMBS lending was around 20% of the market and to them there was no reason why this element would not transfer.

Wilson Lee, then with Lehman Brothers, arrived in the UK in 1996 looking to get a European securitisation business off the ground. He approached British banks, trying to interest them in securitising existing loans from their balance sheets, releasing capital back to them to lend again, giving them an all-important improved return on capital. "'Why would I want to reduce my book? I want to be the biggest lender in the market'," the banks said to Lee, who shut the CMBS shop and began Lehman's principal investing property business.

Back in the US, Lee had been an integral part of CMBS, a market which was again kick-started by RTC workout. The RTC faced a backlash as it sold off loans at a deep discount to investment banks and wealthy individuals, so one way to appease taxpayers was to get better value for money by using securitisation. It took a large number of mainly non-performing loans and packaged them up into a single bond issue. The proceeds of this refinancing were returned to taxpayers while the bonds were serviced from the income and sales from the foreclosure of the loans.

The RTC's backing gave new credibility to CMBS, which until then only had a track record of a few busted deals in the 1980s. It was a resurrection that soon attracted the interest of the US investment banks. Seizing on this new appetite for the bonds from investors, they quickly approached banks to buy their non-performing loans and to fund them using the same securitisation method.

Buying bad loans from banks at 40 cents on the dollar and selling them at 60 cents already worked, but if they could do the same deal by financing that through the fixed income market, this offered an unexpected source of debt to enhance those returns further.

Lehman then took that idea one step further. It took better-quality loans by targeting those that were under-performing rather than in full default, with bond servicing based on interest payments from borrowers rather than sales and foreclosures.

For example, if a bank lent $70 million on a $100 million asset that was in some distress, Lehman would agree to buy that as a sub-performing loan for around $35 million.

It then renegotiated the interest payments with the borrower by splitting the loan into two pieces: payments for a $35 million loan, which matched Lehman's outlay, and then for a second loan of around $15 million for the second piece, wiping $20 million of the debt for the borrower.

When Lehman securitised these loans, the first part formed the basis of the A note, which was securitised, allowing Lehman to receive back its entire purchase price through what the market saw as conservative investment-grade bonds. The idea was that, even in a distressed situation, the underlying building would always sell for at least $35 million and be able to fulfil the bond payments.

The second loan part was the B note, a junior note, a "hope note" as the likelihood of payment was lower and would also be after the A note, but for Lehman was pure profit. In a nascent market, there was little demand for B notes from bond investors so Lehman kept them on its own balance sheet, which was fine as they were "generally at a free or a very very low cost," said Lee. "Lehman became the biggest buyer of sub- and non-performing loans that had those characteristics. We really earmarked those instead of targeting portfolios where we would be required to take the loans through the full foreclosure process."

This bifurcation of performing property loans into an investment grade A piece and a non-investment grade B piece took the market in two new directions in the US. Those, like Lehman, that could source the right large sub-performing loans from which to securitise an investment grade piece, focused on that market. Other investment banks, however, did not have the relationships to access these types of loans so instead started to originate their own. They allocated a pot of money from the company balance sheet and created a factory for smaller loans: £5 million or less, all with a 60% LTV and signed off using the standard documentation.

Once again, the loans were split into investment grade and non-investment grade bonds. The A piece was rated by analysts and split into the different pricing for investment bonds – four classes: AAA, AA, A and BBB – for, at that time, anything up to about 60% of the value of the property backing the loan. Anything below that was the B piece and not considered investment grade, and tended to be held on the loan issuer's books.

This was already a riskier position to take than on sub-performing loans. "At Lehman, if we were buying a loan bifurcating it, and buying it at a good price we didn't mind holding it. It was different if you were originating it because it's your money going out the door, it's your money you are bifurcating and that's still your risk," said Lee.

It was that final step, the purposeful originating of new loans to pool and then securitise, that was the innovation that changed the lending market in the US. This was conduit lending where banks were acting as a pipeline, originating the loans then shifting them off their balance sheet by selling them as CMBS. This was the reason why CMBS were soon 20% of the US market.

## A second entry point for securitisation

In 1997, Morgan Stanley advised Canary Wharf as it raised £500 million of financing with a CMBS issue backed by rents from a pool of buildings on the estate. The deal was the largest ever securitisation in the world and made the potential for the financing tool in Europe difficult to ignore.

When it came to CMBS, the investment banks were not going to let this financing innovation go to waste. But with no appetite from banks freeing up their balance sheets, and too competitive a banking market for conduits, investment banks turned to the listed property companies, which picked it up in record-breaking style.

For Canary Wharf, securitisation enabled it to find a long-term source of debt which also reduced its overall cost of financing. With listed companies still trading at below NAV, there was no possibility of raising capital through the stock market.

The bond market had the potential to absorb these major transactions while bond investors were willing takers in the refinancing of these world-class trophy-style assets.

British Land soon followed suit with the largest single estate securitisation when it refinanced Broadgate estate through a £1.5 billion securitisation. It made sense given the yield gap between property and long-term bonds. The Broadgate buildings were yielding around 7% while the cost of the £1.54 billion raised was in the low 6%.[8]

On the Continent, similar deals took place with portfolios of trophy buildings in Paris. Two early securitisations by Canadian property company SITQ saw a FFr2 billion (EUR305 billion) bond issue backed by rental income from five office buildings, and a second one of FFr1.1 billion (EUR168 billion) secured on three buildings in central Paris.[9]

While the early trophy securitisations helped investment banks stay at the top of the all-important CMBS league tables, they helped to create a market rather than generating high profit margins for the banks.

These were agented transactions, deals where investment banks received fees to advise and arrange for the bonds to be sold into the capital markets. The real money was in conduit lending. This earned fees from borrowers when originating the loan, commission fees from those buying the bonds, as well as the profits from the margin between the interest rates of the loans and the lower bond margin at which they were sold into the market. This all returned the original equity to start that process all over again.

Conduits defied the strong tradition of relationship banking in Europe, but Morgan Stanley was not prepared to let the opportunity go. They hired Lynn Gilbert, the former head of Société Générale's lending team, to investigate setting up Europe's first conduit for the UK market. "All my contemporaries thought I was completely crazy. Why would I give up a nice cosy job running SocGen's loan book and go into the madness of an American bank?" she said.

But these were exciting times for lending, and following Morgan Stanley's lead in the trophy deals, originating to securitise was the next frontier. "We've just had to take a deep breath and make good sensible loans, and then go to the ratings agencies," said Gilbert.

Morgan Stanley allocated £200 million from its balance sheet and European Loan Conduit (ELoC) was born, although Gilbert did not want to make a big deal of this with borrowers. "We didn't try to make it out to be anything really unusual. It was just a loan like everyone else did and we tried to down-play the whole securitisation of the loans. That's the way we manage our balance sheet," said Gilbert.

ELoC also looked totally different to the securitisations seen in the market so far. These were not the trophy buildings that the CMBS bond market currently financed. Instead, they were well-anchored shopping centres in secondary suburban markets, or regional offices and a single property mainly leased to government tenants.

Next, it was for the rating agencies to analyse the portfolio and break it down into the language that sold it to the bond investor community.

The tranching of properties into different investment grades from AAA through to BBB could be a challenging notion for those fixated on the buildings. But CMBS were not about grading buildings, they were about understanding their cashflow and the potential for default, and translating that into layers of risk. For example, a ratings agent might take the view that a portfolio of buildings bought for EUR100 was actually worth EUR70 and this would be the piece they would rate. The rental payments would act as income cover for the bonds but in the event of a default, even in the very worst recession, these buildings would sell for a rock-bottom price. It was that last-gasp fire-sale value – say EUR10 – that formed the AAA, the tranche that could provide the most surety to investors. The next tranche supposed a less dramatic worse-case scenario, allowing for a higher recovery of capital – say EUR20 – to cover the next layer which would then form the AA. This carries on through the tranches, leaving EUR30 unrated (or later rated as non-investment grade pieces from BB to E), which cannot be sold through the bond issue and left with banks to offload through its fixed income desks.

For ELoC, non-investment grade bonds were of no use to Gilbert. The market for CMBS bonds was still immature so they needed to be robust with strong cashflows to assure early bond investors. "It was such a new product, you could just about get away with triple B," said Gilbert.

ELoC 1, a £169 million CMBS backed by 13 loans on UK property, was finally issued in the summer of 1999, with £135 million rated as AAA and the remainder in the three following tranches.

It was a first, as a pool of loans made specifically for securitisation, but it still lost Morgan Stanley money, as a momentary shift in the bond markets widened spreads and swallowed up the profit.

Morgan Stanley was not put off. Within two years, with most of the market still convinced the bank was losing money, ELoC issued a further six CMBS issues to a total of EUR3.9 billion. But it was ELoC 4 that Gilbert saw as the breakthrough, as she lent on six shopping centres in a joint venture between UK property company MEPC and Australian shopping centre owner Westfield. "They pitched us against all the other major lenders and we were the only ones who could write the cheque out because we didn't have to syndicate it," she said. "I felt that as soon as we did that, we had a business as it was a real eye opener for others in the industry."

## Growth and new entrants for lending

By 2004, outstanding property debt held by banks in Europe stood at EUR1 trillion. By country, the top borrowers reflected the dominant lenders in Europe: Germany owed EUR287 billion, the UK EUR238 billion and France EUR124 billion.[10] The CMBS market stood at EUR108 billion.

For the UK, lending was gathering pace. In the second quarter of 2004, the industry saw its highest quarterly rise since records began.[11] The amount outstanding in the UK was two and half times the peak in 1991, with debt playing a much bigger role in individual deals.

After being a cornerstone for the German banks' international expansion, their dominance of the UK market was now ebbing away, down to one-fifth of market share from one-third.[12] The change was partly due to constraints back home. In a deteriorating German economy, they were finally facing up to non-performing loans amassing on their books from excessive lending domestically.

Hypo was first to give itself some more room to breathe on new lending by housing its non-performing loans in a separate division. Now that the US investment banks had cleared the path into the bond markets, the German banks' interest in securitisation also picked up. It could help them lighten balance sheets but also clean them up ahead of Basel II, the revised regulation on minimum capital banks had to hold as a safeguard of their lending.

That said, the troubles at home were not preventing them from some adventurous international lending. Both Eurohypo and Hypo were now lending in the US, and Hypo was opening branches in Hong Kong and Tokyo. In Europe, the Nordics were on the radar screen for both, while Eurohypo was also heading east into Russia.

Taking their place in the UK markets, was the homegrown talent. RBS and HBOS in particular were expanding fast at the expense of the German banks, now commanding around half the market.[13]

Their activities in the UK foretold their ambitions across the Continent. HBOS tended to be more focused on the UK market but after seeing the German banks wounded in their own territory, they joined RBS to do the unthinkable and open up offices in Frankfurt to compete directly[14] in an already overbanked market.

Irish banks also now elbowed their way onto the UK scene, taking a similarly aggressive stance. Lending by Anglo Irish, Allied Irish Bank and Bank of Ireland was relationship banking at its most potent. In Ireland, they had backed their property developer and builder clients who had amassed huge wealth on the back of rising house prices in a booming Celtic Tiger economy.

Now, the banks' clients wanted to be debt-driven buyers in the UK and on the Continent for the same stellar returns as their speculative development at home.

The Irish banks welcomed the segue into other markets, lending both to new and existing clients. By late 2005, Anglo Irish's loan book was at £8.6 billion.[15]

The Irish banks' style was already edgy; activities at home were speculative development loans, land purchases, regularly lending at 100% LTV with limited and fast due diligence to secure deals with regular clients.

They remained confident lenders because they believed that the money they had already helped their mainly private client base earn, would always be a backstop if future lending went awry. Being part of the eurozone was also giving them a false confidence as they now had consistent access to greater sources of capital in the international money markets.

Debt was now a warm bath that the whole industry was enjoying, easing the pain of getting deals across the line, and helping returns stay afloat as prices rose. This felt like a new natural state which resonated with all that was happening in the industry: the opening up of the markets to international players, the unlocking of new countries and sectors to invest in as the size of deals grew. With new products like funds and CMBS, everyone was expected to be more financially innovative.

Then, meeting them over halfway were the banks, with growing levels of debt to lend, as property remained the most promising parts of their business. This was working for everyone. This was the defining trend for the cycle.

In 2005, the level of debt in the European market leapt by 20% to EUR1.2 trillion.[16] Underneath this activity were growing levels of risk. Loans in major markets moved from being priced at 115 basis points and 80% LTV two years earlier, to LTVs of 85% and costs down to 80 basis points.

There was also concern about what banks were lending on. With prime property in short supply and offering declining yields, buyers were shifting wholesale to secondary properties, causing the price premium between the two classes to narrow.

It was now a growing concern that banks would be bullish to lend on the same terms for these properties, disregarding the additional risks attached to secondary markets and poorer property.[17] "If the cashflow looked OK some of the property fundamentals were starting to be overlooked," said one banker.

For every bank that wouldn't do the deal, there were plenty more that did. With the German banks struggling in their home markets, other banks such as HBOS and the Irish banks were willing to step in with mezzanine lending and other higher-risk products.[18]

Refinancing was a growing part of lenders' business as investors took advantage of high pricing to withdraw equity from existing investments to fund new ones, leaving more risks in the hands of the banks. "Investors rode the yield curve," said Max Sinclair, who worked at Eurohypo in this period. "They borrowed at 75–85% then yields went down, rents didn't change and they refinanced at 80% on a new value."

While debt started to embolden all parts of the market, it also sustained an influx of buyers who would max out their leverage by layering on as much debt as possible from senior to mezzanine to base their returns of wafer-thin slices of equity.

These were the likes of long-time investors but also a new breed of speculative entrants to the market who were more financial operators rather than real estate operators. "There were certain people we didn't lend to because it was purely arbitrage," said the banker. "They would ring me up and ask 'what can you lend me', mainly because the bid was based on what we could finance."

This dangerous attitude was starting to pervade all parts of the market, with debt dictating the returns and prices paid for properties rather than the property fundamentals. "Before the market was led by old and conservative pension funds and insurance companies that dominated the central London market, or whatever market you want to talk about, but these institutions moved out of the way because they weren't able to compete," said Joe Valente.

Debt was already prevalent at the prime end of the market, but with prices rising, it was soon going to be a tempting strategy for all buyers and all property. "You saw the debt buyer basically fixing the price," said Valente. "They were the market makers."

# 7

# BLURRING THE LINES

John Carrafiell, head of European Real Estate for Morgan Stanley, was very familiar with Canary Wharf's boardroom.[1] As the company was one of the investment bank's top property clients in Europe, he had advised it on its flotation to the London Stock Exchange in 1999, and been the arranger on its early securitisation deals.

It took him no more than five minutes to cut across the estate from his office at 20 Cabot Square, navigate security at One Canada Square, the estate's central skyscraper, to take the lift to the company's headquarters high at the top of the tower. But in late 2003, that short journey might have felt longer as he entered the boardroom to confirm to Canary Wharf's Chairman Paul Reichmann that he intended to buy the company on behalf of Morgan Stanley's real estate fund.

Morgan Stanley's ambition to buy the east London estate was a world away from the investment bank's starting point of broken-down disasters of non-performing loans in France, or the under-valued buildings trapped on corporates' balance sheets or hidden beneath listed property companies. This was 1.3 million sq m of prime, gleaming offices and shops, finally finding momentum after surviving bankruptcy in the 1990s.

One Canada Square, the tallest building in London, was recognisable by its pyramid roof, clad with stainless steel as a nod to Britain's heritage as an industrial nation. Now, it was the centrepiece of the estate's financial prowess, flanked by boxy buildings boasting efficient floorspace, making them the favourite of major financial institutions such as Credit Suisse First Boston (CSFB), Citigroup, HSBC and Morgan Stanley.

A new area at Heron Quays was just coming to life around the Underground's new Jubilee line extension, a much-needed transport artery. Lehman Brothers was due to move into its new European headquarters in the same area at 25 Upper Bank Street, while an underground shopping plaza was bringing much-needed relief to the sequestered office workers.

It is true, this picture of health did belie some financial strain. Canary Wharf Group was finding it difficult to breathe under its debt levels and profits had slumped against the backdrop of a lacklustre London office lettings market. The weakness was apparent in its dwindling share price, particularly after the company admitted that it had failed to disclose promises to take back more than 60,000 sq m of unused space for up to ten years as part of negotiations to secure mammoth lettings to companies including Lehman Brothers.

However, any distress signals emanating from Canary Wharf were a world away from the ones opportunity funds were used to picking up on. MSREF's plan to bid for the London estate was a sure sign that as the industry moved into 2004 and beyond, the landscape for the opportunity funds was changing. Whether it was trophy assets like Canary Wharf or portfolios of secondary properties, the funds now needed to get more creative to get capital deployed. The markets were expensive and fewer of their earlier strategies now differentiated them from other players in the market.

They could still use scale to their advantage. If they couldn't buy big with rare opportunities such as Canary Wharf, they would aggregate, building their own specialist platforms and portfolios. This would see the playbook changing with them taking on other types of risk, whether that was operational or property development, or more risk when it came to their favourite, debt. The funds still had access to extraordinary levels of debt and, in the end, whatever they did, leverage was always at its core.

"You went from being a distressed player to being a growth story player, so you were expecting rental growth; you were thinking at the margin, you could improve the cash flow. But the real story was that debt was so available and cheap that you could get those returns through leverage," said David Brush.

This last part of the cycle, the grandest phase of investing yet, would also reveal some cultural differences between the players as the investment banks – Goldman Sachs and Morgan Stanley, in particular – pulled away from the pack, in both the aggressive nature of their investing and use of leverage, as well as blurring the line of their own interests – including personal gain as individuals – versus that of their investors.

## A one-time opportunity extended

It was 2003 and no one in Europe thought to ask opportunity funds what they were still doing there. All these funds, particularly the financial players – MSREF, Goldman Sachs' Whitehall and Lehman Brothers – were created to be anti-cyclical; higher-return investing was a rare sport, to be practised only at specific times in the cycle.

When the market had needed it most, the funds had profited from plugging it with equity, but these dislocations were by definition cyclical. Now, in Europe, that early phase of 20% plus return investing was over. Equity was the least of Europe's problems; it was a fully lubricated machine as it tied down a healthy EUR67.5 billion of deals in 2003.[2]

Competition for deals was already pushing up pricing, affecting returns; opportunity funds launched at the back end of the 1990s were projecting between 15%
and 18% returns, not the 40% seen with the earlier vintages.[3]

One of the main problems was deal flow. The type of deals that were standard
fare for the opportunity funds were now proving problematic. Major corporate
spin-offs fell out of bed due to sellers' unrealistic expectations, angering bidders and
wasting millions in dead-deal costs. In the listed markets, narrowing discounts to
NAVs meant fewer opportunities to arbitrage between the two markets by taking
companies private.

German supermarket chain Metro crushed the ambitions of Morgan Stanley,
Goldman Sachs, Blackstone and Merrill Lynch by blowing hot and cold twice on
the sale of a EUR3 billion portfolio,[4] sucking any residual confidence out of the
market for these spin-offs.

The same happened in Spain with a EUR340 million sale and leaseback deal
with national telecoms company Telefónica, costing RREEF (formerly Deutsche
Bank) and Spanish property company Metrovacesa EUR1 million in sunk costs
and six months on a wasted deal.[5] RREEF together with French banking group
CDC finally succeeded in buying a EUR1.4 billion portfolio from ENEL, but
only after the Italian energy company disappointed another bidder and shrank the
original EUR2 billion package.[6]

Not only was pricing increasingly out of sync, but corporates were also beginning to realise that what the investors were offering was at odds with their own
occupational needs. Fewer corporates lined up to sell and those that did were
much more particular about what they would put on the market and under what
terms. Even the banking community was becoming wary. These deals relied too
much on a single tenant credit and there were concerns about the value of the
underlying property as leases ran out.

These early corporate spin-offs were ground-breaking, but as the opportunity
funds tried to keep up the tempo from the early days in the US, they were less
about innovation and more about survival.

The staying power of the opportunity funds was already unexpected. Even
when the funds were first set up in the US, they were thought of as a one-time
pass, a strategy that would last only as long as the RTC. "I know in the first generation of those funds, people thought they were a one-time exercise. Literally a
one-time exercise," said Peter Linneman, principal of Linneman Associates, and
Albert Sussman Emeritus Professor of Real Estate, Finance, and Public Policy at
the Wharton School of Business at the University of Pennsylvania.

As the distressed deals slowed down in the US, Europe gave the funds a second
lease of life. Now, even on this side of the Atlantic there was a crowd. It was
becoming apparent that for this type of investing, there was room for just a
handful of fund managers with the right mix of market knowledge and money.
Yet, they sat in their own outer layer of the property atmosphere with around
EUR23 billion of dry powder between them to invest in new opportunities in
Europe alone.[7]

Even as they arrived in Paris, they had baggage. Such was the success of the early investments in the US that they were under extraordinary pressure to replicate that by investing big and investing fast. "They just were under a lot of pressure to shove money out of the door, and they adhered to orders," said Roger Barris, who worked for Goldman Sachs and Merrill Lynch in this period.

Back in the US, the first funds netted investors returns far above the 20% promised, setting a level of expectation that was going to be difficult to disappoint, but also giving fund managers a new lucrative line of income that was impressing Wall Street. "The reality is that these guys were raising very large funds because collecting investment management fees is attractive for the bottom line of their publicly quoted parents," said Barris.

This resulted in rounds of US opportunity funds launched in quick succession, immediately oversubscribed by repeat investors and newcomers. The first White-hall fund in late 1991 raised just US$166 million from investors. The second, a year later, climbed to $790 million. When Whitehall Street Real Estate VII closed in 1999 – its first fund to target Europe – it had $1.86 billion of equity.[8] Layering debt of up to 85% into future deals, the upper estimate of its spending power was $12 billion.

At Morgan Stanley, its first fund, MSREF I, was $585 million of equity and, just prior to its full liquidation, it had given investors a 48.5% internal rate of return.[9] That was followed with a second fund with double the equity at $1.2 billion, and a third – split into domestic and international pots – of a combined $2.3 billion.

Success reversed the control, with fund managers now needing to find the opportunities to drive the funds, rather than the other way round. "If you've raised a large amount of money, you know that you are not going to be able to raise the next fund with the fees until you've invested this money," said Ian Marcus, who worked for CSFB at this time. "Just put the money to work, just put it to work."

But that growth also meant that funds also had to contend with whether the fund structure could withstand this newfound longevity. The fund model wasn't even originally created for property and was instead repurposed. "The early movers from investment banks did a 'copy paste for their non-property private equity (merchant banking) arms and applied it to property. Same documents. Same fees. Same investors. At least initially," said Struan Robertson.

To merchant banking, a 20% IRR may have meant something but to property it was arbitrary, and only encouraged assumptions in Excel spreadsheets to make sure that 20% worked on paper.

Linneman said he once asked two iconic investors – one in hedge funds and one a private property investor – what their best return was on a sustained basis over a decade or more. "They both had 12%. And I remember thinking, if these legendary guys on a sustained investment basis can do 12 and that was when interest rates were 4 and 5%, how does everyone think they can do 20?"

For now, 20% appeared easy enough to the fund managers when they could see 30% to 40% in the rear-view mirror from early funds, but as they moved forward into a different part of the cycle, there was a danger that their strategy would become more of a scattergun venture-capital approach. "I'm going to pursue things that give me a 20% return on paper but don't expect a 20, you are either going to get a 44 or minus 6," said Linneman.

The growth in fund size was also distorting the fee model. Managers were paid a management fee – typically a percentage of capital raised or invested – which was expected to cover employees and the basics with a cut of the profits if they generated high returns for investors. This worked fine for the early, smaller funds, but when managers started to get to $10 billion under management with, say, a 1.5% fee, that was $150 million in fees. "You've got to have a lot of secretaries to not have someone making five or ten or fifteen million dollars," said Linneman, and that's before the performance fee. "Of course, they still want the promote but the balance starts changing."

The size of the management fee shifted the emphasis to getting money raised and deployed rather than investing it well. Any performance fee, which was part of the incentive package, was much less relevant. Teams were earning a salary and the bonuses were based on growing assets under management (AUM), which aligned with the banks' financial reporting targets. "The capital markets did not value individual performance, they valued the fundraising machine," said Christophe de Taurines, who worked for CSFB and LaSalle Investment Management in this period.

Funds also became more expensive to run. To begin with, they thrived with small teams of bright people on the ground but with success this evolved into a more formal infrastructure that needed feeding. "The trouble with private equity vehicles, is that you are like a hamster on a wheel," said Marcus. "It's only those with multiple funds that could really make it happen, so you get on that treadmill and go and raise the next fund and then you dilute the cost of that fundraising over several funds."

By the early 2000s, the US opportunity funds were too embedded in Europe to pull back. "The hardest thing that I've probably seen in all of this is dismantling the machine. It is a very hard thing to do. One, you've put it together, you are proud and it's working. Two, it's expensive to dismantle and three, in the back of your mind, you tell yourself, well, the opportunity will come again," said Linneman.

In the background all this time were the investors – from US pension funds and insurance companies – who were content to be taken on this ride. Those committing to the first rounds were delighted to be early adopters, quickly taking up future spaces in what would soon become oversubscribed funds.

The relationship that the opportunity funds had with limited partnership investors placed much more emphasis on limited than partner. Once the capital was handed over, the opportunity funds were like black boxes to investors as they had full discretion over all investments and strategies.

Investors had little option other than to back whatever the banks came up with next. "The investors never made the choice to be in Europe. The manager did. Because if you invested in Whitehall in those days or MSREF or any of those guys, you were investing in a global fund, that's the way it got done," said Marc Mogull.

So far, that confidence had been rewarded: whether it was corporate spin-offs in Europe, or non-performing loans in Asia, investors allocated larger slices of capital as they learnt to trust the teams. "Each time you were like 'really, really, really?'" said Joseph Stecher, who worked for General Motors Asset Management and Goldman Sachs in this period. "What I am saying is that we were comfortable with being uncomfortable with their latest ideas."

## The battle for Canary Wharf

As the air seeped out of the corporate spin-offs, the battle for Canary Wharf only became more inflated. The sales process was a fiasco. Six months since MSREF first declared its interest, an auction process led by Canary Wharf's advisers had failed to see a recommended bid on the table.

Bidders were mired down in tactics to outmanoeuvre those now entering the field. Morgan Stanley quickly pacified an approach by Goldman Sachs by bring them into its consortium. Brascan, a minority shareholder, courted 73-year-old Canary Wharf founder Paul Reichmann for a partnership before setting out on its own. Undeterred, Reichmann worked up his own bid, ultimately failing when banks declined to back the man who had already taken the company to bankruptcy once before.

Despite it being a friendly offer, bringing along Canary Wharf's current CEO and finance director, MSREF's first bid of 255p, valuing the company at £1.5 billion, was roundly rejected. Analysts were astonished at the low-ball price, 23% below the last reported NAV as well as their estimates of the break-up value of 295p. They questioned whether it was right to sell a "highly geared, cyclical company at the bottom of the property cycle to a real estate opportunity fund."[10]

In early 2004, MSREF's bid was now up to 275p as a counter-bid to a hostile 270p offer[11] from Brascan. MSREF outlined its future plans for the estate. It would sell off large chunks of the estate and refinance other parts to raise money, with the company continuing to manage a slimmed-down portfolio of buildings.[12] One Morgan Stanley director at the time denied it was a "smash and grab"[13] raid, but analysts were quick to point out that Brascan's outlook was more long term.[14]

Soon, the protracted fight started to look personal. MSREF had opted for a scheme of arrangement offer, which set the winning line at 75% of shareholders. This gave it full control of decision-making on issues such as financing without interference from minority shareholders. Brascan, meanwhile, had amassed 21% of shares, including Reichmann's, who was "irrevocably committed to vote against"[15] the Morgan Stanley-led bid.

By now, the back and forth of bids and counter-bids had become fractious and prolonged, taking more than a year. In 2003 alone, managing the sale cost Canary Wharf Group £10.7 million in advisers' fees.[16]

The Takeover Panel, which supervises and regulates takeovers in the listed sector, stepped in to end the battle with a quick-fire auction. An 11th-hour 295p a share bid from Morgan Stanley sealed it, but not without a concession. At the last minute, it switched its offer to a conventional one requiring just 50% of the votes. The deal valued the company at £4.7 billion including its £3 billion of debt. The price was £200 million more than when the company was first put in play.

The new company, Songbird Estates, would own the company through an Alternative Investment Market (AIM) listed company and brought together an impressive list of shareholders including MSREF, private investor Simon Glick, British Land, Goldman Sachs' Whitehall 2001 fund, and a second Morgan Stanley fund, Real Estate Special Situations Fund II.

Losing bidder Brascan was not put off. It bought up shares including those of Reichmann, building up its holding of 26.5%,[17] giving it an ongoing say in major decisions in the company.

The Canary Wharf takeover grazed the bottom of the London office market on its way up to become one of the most successful real estate transactions of the decade. Within three years, the consortium of investors was sitting on paper profits including dividends of more than £1 billion. By late 2006, their combined equity of £616 million invested in July 2004 was worth £1.7 billion. Songbird saw its shares soar from 100p to 241p.[18]

For Morgan Stanley, the deal not only displayed its prowess on the European property scene but also its increasing audacity to seal deals that blurred the lines between its responsibilities to its fund investors and its own bottom line.

If this large trophy deal was a great deal for MSREF, it was an even better one for Morgan Stanley, as an investment bank. The Canary Wharf deal supported the bank's growing agenda to use the transaction to generate fees for the bank. "There was quite a lot of commentary that the investment bank-affiliated funds were quite good at generating revenue from those funds. They would do financing, acquisition fees, corporate advisory fees. It was a captive pool of money which those investment banks were able to tap into as captive clients," said one manager.

The market was astonished at the temerity of Morgan Stanley to shift quite so publicly from gamekeeper to poacher with such a big client. "I said to John Car-rafiell during that time, 'I get it now, you advise a company, float it and then you take it private. Perfect, you keep it all in the family'," said one rival adviser.

With such a drawn-out bidding process, the total fees going to advisers were estimated to be £200 million divided between 40 professional firms including lawyers, accountants and nine banks, making it one of the most expensive bidding battles in UK history.[19]

As a listed company, Canary Wharf Group's finances were in the public domain but it was Morgan Stanley's insight and intimate knowledge of the estate as its long-time adviser that gave it an unsavoury advantage as the bid was made. "It was an opportunity that they controlled as they were the adviser," said another banker. "They saw an opportunity through that control position to come in and take the thing out. Simple as that."

Already from a property perspective, Canary Wharf provided the MSREF with many bites of the cherry: buying it from the listed market at a discount to the value of its assets, and getting prime property with a recognised management. It was also an opportunity to pile on more debt and generate returns on the sliver of equity in the deal, and with Canary Wharf's land holdings, it was a development play. "You can argue the ownership was fractious, it was big assets, looks good in the photos, has strategic importance to London and a massive ticket size. It ticks a lot of boxes for a fund," said one former private equity manager.

In addition, Morgan Stanley was the adviser to MSREF and the lead adviser to Songbird for the takeover. It then arranged £737 million of new debt to replace the mezzanine and senior. The new facilities, which aimed to be more flexible and cost effective, were provided by Hypo Real Estate and Morgan Stanley.[20]

Ironically, Morgan Stanley also advised on unwinding the pioneering securitisations it put together for Canary Wharf as its client in 1997 and 2000, arranging alternative financing including a £750 million loan from RBS.[21] "I'm not sure who paid John Carrafiell, the investment bank or the fund, but there would have been those kinds of conversations," said one private equity manager.

Carrafiell's responsibilities straddled all Morgan Stanley's main real estate activities – the lending side, advisory work as well as the investing on behalf of clients through the MSREF series. Morgan Stanley saw this as a benefit for its clients, a unified division that could offer a full range of solutions. For the market, it raised issues of conflicts. "Were there conflicts? No question but all investment banks have conflicts and the key question was how you managed them," said one investment banker.

There had been the same murmurings five years earlier when Morgan Stanley listed Italian company UNIM, only for MSREF to buy properties from the company that took it private months later.

One banker saw the Canary Wharf deal as a shift in power within Morgan Stanley's real estate division. Carrafiell's rise to head of Europe, now gave the advisory side more clout. Under previous leadership, Milano Centrale's acquisition of UNIM and subsequent sale of properties to MSREF had undone the work of the investment banking division. "Five years later it was 180 degrees the other way. Private equity guys were fulfilling capital needs for investment banks' clients," another banker said.

This drive for fees was not only happening at Morgan Stanley but its ambitions to be a full-scale diversified real estate platform, as well as its tendency to be much more high profile in the market, meant its activities were more obvious than most. "A lot of landmark transactions of that period didn't work out so great for the LPs but they worked out pretty good for the sponsors," said one investment banker.

## The power of leverage

Over at Goldman Sachs, as it scaled up its investing and the size of its deals at this stage of the cycle, it became as ambitious with its use of leverage. Compared with Morgan Stanley, Goldman was seen much more as traders. "The Goldman Sachs guys didn't care about the downside case because they had no incentive in the deal and it was a just a trade, an opportunity. They weren't building a business," said one fund manager.

While there was continuity at the top level, below it was a constant stream of analysts and associates. "Goldman Sachs had a scorched earth view about their vendor relationships, everyone hated working with them. They were rough, hard at the edges," said the fund manager.

Straightforward deals with stacks of debt were classic Goldman plays. Competitors often stepped back if Goldman Sachs was in the running with access to high levels of financing.

It first showed that it could scale up in Paris in 2004, when its Whitehall fund bought 50% ownership of Coeur Défense. By now, the landmark building was 99% let and in Unibail's eyes, its top price achieved. The office complex that had cost EUR665 million to build in 2001 was now valued at EUR1.345 billion.

Under Bressler's leadership – himself, a former banker with experience working in the US – Unibail had already released much of the capital from the building through a EUR820 million securitisation and EUR505 million of subordinated debt.

By Unibail's calculations the level of equity left in the asset was around EUR30 million. The CMBS issue stayed in place while Unibail sold most of the EUR550 million subordinated debt to the lending division at Goldman Sachs International and Hypo Real Estate.

This was a mammoth deal for the market but the equity amounts were tiny. Whitehall was likely to have handed over a maximum equity payment of EUR15.1 million.[22] "It was just typical Goldman Sachs. A great deal for them," said one banker. "They were trying to turn every deal into that deal."

This was not about the property but the high quality helped. This was about the multiplier effect on the equity of the staggering levels of debt, easily clearing the 20% return hurdle the fund required on its small amount of equity.

Goldman followed this by taking its format to one of the largest deals in European property history. The EUR4.5 billion Karstadt sale and leaseback was secured close to the peak of this cycle's 2006, showcasing the bank's ability to max out debt.

It was an awkward entrée to the deal for Goldman; the investment bank was advising German retailer Karstadt on its restructuring, and circumvented a planned auction to become the preferred buyer. Rothschild stepped into its place as adviser,[23] but still, with its work on the deal, Goldman racked up "obscene" fees of EUR32 million.[24]

The portfolio of around 157 mainly retail stores from Karstadt's high street brands valued at EUR4.5 billion, transferred to a joint venture owned by Goldman Sachs' Whitehall fund and the retailer itself. Karstadt was paid EUR3.7 billion with a further EUR800 million to come if the properties went up in value.

Goldman structured the deal which included around EUR3.4 billion of debt in three different loans, all secured on EUR259 million of rent a year. Whitehall was thought to have contributed just EUR100 million of equity for its 51% stake in the EUR4.5 billion joint venture.[25]

At the time, observers thought the portfolio could be worth as much as EUR6 billion but this depended on the health of Karstadt. The restructuring was to help recapitalise the retailer as a response to its losses in 2004 of EUR1.6 billion. Even though it had already shed jobs and stores, its future was still uncertain.

With deals the size of Canary Wharf or Karstadt few and far between, the opportunity funds were having to be more creative about where and how to source these later-cycle opportunities.

The easiest way to get scale was to build the portfolios themselves, by aggregating property from several deals.

Once again, Morgan Stanley was at the forefront, now taking on much more property risk alongside its financing risk. At the start of 2006, it paid EUR479 million for the shopping centre development arm of Dutch developer AM, as part of a EUR995 million joint venture with Dutch developer BAM. The deal was struck at 32% above the company's closing share price.[26]

From ten years of investing in Europe, the purchase propelled the investment bank to a new status of being the continent's largest developer with a pipeline of projects with a value of EUR20 billion.[27]

AM, renamed Multi Developments, formed a central plank in this pipeline with a programme of EUR9 billion and a presence in mature European markets such as the Netherlands, France, Germany, Spain and Italy, as well as growth markets such as Turkey and Central Europe.

MSREF was now directing around 40% of its capital for Europe into development, adding to this a EUR1 billion joint venture with its long-term Italian partner Pirelli Real Estate, as well as a £300 million portfolio of 21 development sites, more than half of which were in the City of London.

The rationale, according to Carrafiell at the time, was that occupier demand was on the horizon in key markets with rents soon to rise in central London. Even though MSREF had no roots in development, Carrafiell said its size would enable it to commit resources to "new markets, assets classes and structures".[28]

It then looked to do the same with hotels, with plans to build up a EUR2 billion portfolio after buying seven InterContinentall hotels for EUR624 million later in 2006. That brought its ownership to 14 hotels and 4,000 rooms across Europe, as it looked to make money by expanding and managing them better.[29]

RREEF was also getting creative in a bid to secure large portfolios with two major retail acquisitions. Across the two deals, both in partnership with Italian retail group Borletti, it paid EUR888 million for 166 stores, including 18 department stores, from Italian retailer Rinascente, and then a year later EUR1.1 billion for 17 Printemps department stores. In both cases, the difference was that the full retail operations came along with the deal. "I put zero on the operations and said I'm getting EUR1.5 billion of property value for EUR1.1 billion. Then if I get anything out of the operations, that's great," said Brush.

## Looking for the exit

Creating such mammoth portfolios was all very well but when you were already the biggest players in the market, finding a wholesale buyer later down the line was hard. This was one of the reasons why from the early days US players were so keen to promote an enlarged European listed market. It was one of the only places that could offer the liquidity needed to exit. Now, some of the planets for the European REIT market were starting to align.

Without warning in early 2003, the French government ordained the Sociétés d'Investissements Immobiliers Cotées (SIICs), a tax transparent vehicle largely inspired by US REITs, and the first true equivalent in Europe. It had all the beneficial tax status for French companies, and with no restrictions on foreign ownership, the French listed sector was expected to attract more international capital.[30]

There were also serious moves to create similar vehicles in the UK and Germany, as well as in Italy, Sweden and Finland.

Unexpectedly, the SIICs attracted a wave of international capital paying high prices from Spain. Metrovacesa paid EUR5.5 billion for Gecina, a French company twice its size, for 9% above NAV, transforming it into the largest listed company in Europe.[31] Colonial paid EUR759 million to get a controlling stake in SFL.[32]

This cross-border consolidation should have been the European dream but analysts derided it as "growth for growth's sake". And while the Spanish companies had lofty ambitions, their plans were driven as much by the SIICs giving them a tax-efficient route to the French market[33] and a way to reduce their tax bill in Spain.

Morgan Stanley had already seen the potential of the SIIC launch pad, selling its 20% stake in, Foncière des Régions, giving the French listed property company (and soon-to-be SIIC) full control of an EUR850 million portfolio the two companies had bought jointly from utilities companies France Télécom and EDF. It was a rare example of an exit through the listed markets for the opportunity funds.

Morgan Stanley also started to back a second company, Mines de la Lucette, providing it with most of its EUR328 million of fresh equity to buy property across France.[34] However, GE's acquisition of listed company Sophia for EUR1.5 billion provided the usual warning that where there was growth in the public markets, there was also opportunity.[35]

The arrival of these different REIT-style regimes was a welcome development for the opportunity funds, and would provide the perfect exit for the large portfolios it was aggregating. While it had taken Multi private, Morgan Stanley said it did not rule out another listing. For its hotel programme, it wanted to look at floating the business when it reached EUR2 billion. With the listed market now going in their favour, it gave the US opportunity funds the confidence to move to Germany for their next and final major macroeconomic play.

# 8

# CRISIS HITS GERMANY

After German open-ended fund CGI paid out EUR2 billion for the White City shopping centre development, it stood by as the billionaire Reuben brothers and Nina Wang tried to throw their weight around in the landmark project already underway in West London.

CGI owned 100% of the rights and what the billionaires had in their hands was an option to later buy back 50%, a position that was the opposite of their natural inclination to be in charge. CGI was unmoved.

With this single deal in the summer of 2004, the German open-ended funds showed that the pinnacle of their powers had been reached.

The billionaires tired and sold their option to Westfield, Australia's leading shopping centre developer, desperate to break into the European market.

With CGI still in the driving seat, it negotiated a EUR400 million contribution to the development and the right to take home a preferred 6% yield on income. Today, ten years after the centre opened with the partnership proving successful, the shopping centre at White City, Westfield London, is still one of the fund's best-performing assets.

CGI's success at White City reinforced the aura of the German open-ended funds. By 2002, the Germans had overtaken the Americans as the largest group of investors in European commercial property. They – the majority was the open-ended funds – were responsible for EUR9.3 billion or 29% of all cross-border deals, the US just behind with EUR8.6 billion.[1]

The German open-ended funds also accounted for six of the top ten investors in Europe. CGI topped the rankings by spending more than EUR2 billion of capital in 2002,[2] just two years before White City, a single deal of the same size. But this high point was also the start of an increasing contradiction between the funds' domination outside Germany and growing problems at home.

Just one month after CGI signed its record deal, Michael Koch, a managing director at competitor fund manager Deka (formerly known as Despa), was sacked over "irregularities".

He was later found guilty of corruption by a Frankfurt court, admitting that he had accepted EUR470,000 in bribes in exchange for helping developers and architects win building contracts across three cities in Germany.[3]

Koch was the most public figure in a real estate scandal that engulfed the German open-ended funds between 2004 and 2006. At the height of the investigation, Wolfgang Schaupensteiner, the Frankfurt public prosecutor, said he was interested in speaking to up to 80 people, citing bribery, breach of fiduciary duty, money laundering and deception as well as tax evasion. Soon after Koch's sacking, the entire management of Deka was forced out of their jobs.[4]

For Schaupensteiner, the investigation's scope was linked to 30 to 40 properties worth between EUR3 billion and EUR4 billion in total. "It goes all the way to London," he told *EuroProperty* at the time. "After all, Europe is small."[5]

As a formula that had worked for 40 years, the continued success of the German open-ended funds was unsurprising. For the last decade, deal by deal, their investing in Europe outsmarted the markets and drew admiration as they snapped up buildings as though to the beat of a drum.

The German public was hypnotised by the rhythm, taking the inflows of cash to a whole new level. In 2002, they more than doubled. Month after month, over EUR1 billion of cash passed across the counters of banks, providing the funds with EUR14.9 billion of cash to invest compared to EUR7.3 billion the year before.[6]

For the open-ended funds, size denoted success but it also brought about its own challenges. Already through the 1990s, funds grew at an average rate of more than 20% a year.[7] Now, in the four years from 2000, they ballooned from EUR48 billion to EUR85 billion.[8]

There was no let-up for managers. Speed to market was vital. Regulations dictated that they could hold no more than 49% of their worth in cash, driving fund managers to convert money into property. In 2003, many funds were running cash ratios of 20%; some were at dangerously high levels of 40% and more.[9]

The German fund managers' demeanour stayed calm, but at times the stress started to show. "I remember having conversations with open-ended fund managers saying, 'I've got this massive capital pouring in, can you go and find us something to buy'," said Robert Orr. "I exaggerate the point but it was a plea for help because the timing and quantum of capital inflows weren't always controlled by the fund managers."

In most cases, fund managers had little say over the timing and speed of the inflows. Occasionally, commissions for sales teams were lowered to steer them away from bulging funds but in general the tap stayed on. "We were on this lifeblood of inflow and outflow but that was driven by global financial markets, that was driven by marketing strategies by the bank but not by real estate fundamentals," said Wenzel Hoberg.

The banks also did everything to stimulate the growth of the funds. In the early 2000s, alongside the German public, they encouraged inflows from institutions and other investment products such as funds of funds. It was an efficient way to bring in capital, and fund managers waived the standard 5% subscription fee to encourage them. However, ease of entry also meant ease of exit.

Success naturally attracted new entrants. By 2003, the 13 original funds swelled to 22, both soaking up growing inflows and stimulating more. New funds took advantage of loosened restrictions on where they could invest, often breaking completely free of Germany.

DB Real Estate's[10] global fund launched in 2000,[11] quickly followed by similar funds from Deka and DEGI in 2003.[12] The DEGI fund was typical, targeting prime assets in the main European capitals outside Germany first, and then a second phase of investing in North America and Asia, followed by Latin America and Australasia.

Bolder still was CGI's global fund. It restricted its European exposure to 20% with half targeted at North America and the rest in Asia-Pacific. CGI's global fund attracted more than EUR230 million in its first five days, underscoring the popularity of the better-performing international markets.[13]

A second wave of funds came from new entrants, established financial institutions with existing banking distribution networks such as French insurer AXA, but also unaffiliated players such as independent fund manager KanAm. "We were looking to reinvent the industry," said Michael Birnbaum, head of communications at KanAm Grund Group. "This was the only open-ended fund without a bank in the background. This was plainly people only doing real estate, no banking."

Operating through a network of independent salespeople rather than those affiliated to each fund's banking parent, it also planned to ditch the buy-and-hold model. "The German word for real estate is *immobilien* which means immobile, and there is some truth to that. But the attitude we had was that we took real estate as something to put money in, to get more money out. If it's done, you sell it again. We started to think about doing it more 'mobile' than the others," said Birnbaum.

With a clean sheet and no German investments, KanAm challenged the sector on its performance track record. Without Germany, it offered a reduced tax exposure for investors; taxes were paid on the fund level in locations outside Germany but individuals would then not be taxed again on the income at home. "The industry didn't like it so much because then the youngsters came in. They were doing their stuff at 4% before taxes and then we are doing it 5% to 6% after taxes," said Birnbaum.

KanAm hit a nerve and its unaffiliated formula saw cash redirected its way. From 2002 to 2004, it had periods when it was receiving EUR400 million a month from investors. Building on its success, in 2003 its first US dollar-denominated fund received EUR8 million in the first three days.[14]

## Deeper in London and Paris

On the ground, the open-ended fund managers met swelling inflows with renewed effort. A new urgency flowed through their latest acquisitions as the pace of investing picked up. The number of deals grew, as did the size as they plumbed established markets and moved into the further reaches of Europe and beyond on the hunt for the right stock with higher yields.

They dived deeper into their favoured markets of Paris and London, buoyed up by the relative safety of Europe's two largest and most tradable property cities. Paris remained an enduring favourite from the moment they first arrived in 1998. Five years later, the sheer extent of the early 1990s crisis left a city still further behind in the cycle compared to its European peers.

Fund managers narrowed in on prime, fully let buildings, stoking the investment market by snapping up properties in larger and larger portfolio buys. Mid-way through 2002, agents estimated that EUR6 billion to EUR7 billion of German open-ended fund cash was poised over the city despite an already frenetic first half of 2002.[15]

CGI continued to lead by example. It splashed out EUR460 million in the largest deal by an open-ended fund in the city so far. The sellers – Goldman Sachs' Whitehall fund and its local partner, Shaftesbury International – originally bought one of the buildings for EUR53.4 million in 1998. Now, four years later, a redeveloped 7 Place d'Iéna achieved a record rent from its new tenant, before being sold to CGI for EUR140 million.[16]

As the funds radiated to the outside of Paris, they ploughed more money into development. Paris was rejuvenating its stock with crops of new offices on the *Périphérique*, the main ring road around the city. As new business hubs, the locations were obscure – Issy-les-Moulineaux, Saint-Denis and Neuilly – but this was a new generation of well-connected buildings with large floorplates, enabling companies to consolidate and upgrade into new headquarters.

Across the Channel, the funds turned to investments outside London to find higher yield combined with security of the UK's long leases. In Birmingham, iii Fonds rocked the market in 2003 by beating out rival German open-ended funds to pay more than the £45 million in its largest deal that year.[17] The next year, DB Real Estate debuted in Manchester, paying EUR75 million for 100 Barbirolli Square.[18]

By now, each UK property agent had aligned itself to one or two funds, and major buildings up for sale in the capital were distributed like a croupier dealing cards. "The inflows were so huge, the lot size was important. There was plenty of property companies and domestic companies below the £100 million threshold but as you got above £100 million, that was where the Germans were the best buyers," said Hoberg.

All the German open-ended funds picked up the pace of investment in newer territories: Sweden, Copenhagen and Belgium, as well as more emerging locations such as Spain, Italy, and Central Europe. While these were seen as higher-risk countries, this was mitigated by membership of the euro or planned accession to the European Union. Tactically, they provided higher yields than London and Paris but high-quality buildings with international tenants could provide better returns. "The question of yield was the most important thing, and the markets were booming so that was positive. I wouldn't say we went to Madrid because there it was low risk or high risk or sustainable yields. It was a mix of there being more money and we got more pressure from our parent company that we had to become more international," said Thomas Beyerle, who worked for DEGI at the time.

Spain quickly became a new favourite, as the open-ended funds redoubled their buying efforts, fending off experienced local buyers to dominate the fierce end of the prime market. Early deals were around the EUR50 million mark but now they had the confidence to make major plays. Deka beat out two domestic buyers to pay EUR115 million for El Triangle retail and office complex in Barcelona.[19]

## Troubles at home

Outside Germany, the funds appeared unstoppable but this was misdirection, diverting the attention from its performance at home. In their own back garden, their domestic property market was in a parlous state. The dotcom bust had sent a one-off cold blast of air across most major European economies, and coupled with the ongoing difficulties of reunification, had settled as a chill in Germany.

Recession was washing through the economy and into the property markets. For the open-ended funds, it exposed the glaring underperformance of their ageing domestic portfolio, magnifying the general policy of rarely selling. Over in the German banks, bad loans from a recent development boom were piling up. These took their place on top of loans to developers that were caught up in the enthusiasm and subsidies to build in East Germany.

Out of the 22 funds, half produced returns below 3% in 2003, and four were below 2%. In early 2004, it was predicted that up to three funds would produce negative returns that year, the first time in the funds' 46-year history.[20]

At first, even hard evidence that Germany was in recession did little to shake the confidence of those putting their money into the funds. Their commitment was based on previous performance and, for the last 60 years, property values had increased every single year, not even dropping in the face of falling rents and occupancy levels.

Fund performance was suffering because of their exposure to the increasingly ageing stock of German properties. The long-term legacy of buying and holding buildings combined with the fundamentals of the way Germans valued property was catching up with them.

The funds' focus on sustainable value, which smoothed out returns over the long term, treated factors that normally dented value in other market such as periods of vacancy, as temporary setbacks.

This encouraged them to be holders rather than sellers of property, renewing through light refurbishment but not through trading. By now, the quality of the ageing stock meant its future viability was exacerbated by the recession.

The problems began in the second half of 2003, as a sharp slowdown in inflows derailed what was on track to be a record year. In the last six months of the year, funds took in just EUR1.49 billion compared with EUR12.2 billion in the first half.[21]

Cash fell away from German-focused funds including those managed by Deka, DB Real Estate, and WestInvest.[22] Most dramatically, one of DEGI's, a EUR6.9 billion fund which was producing an annual return of 2%, lost EUR1.1 billion in the month of December alone.

A few months later, in a market already agitated by poor returns, the corruption scandal broke. The news was a sucker punch to the hundreds of thousands of Germans who entrusted their personal wealth to the sector.

It shattered the fund's image. Their managers could rationalise under-performance in challenging markets but they could not manage headlines, or explain away a crisis of confidence based on corruption. The possibility of losing all their money went to the very heart of investors' fears. Without missing a beat, they acted with their feet.

By the end of 2004, net inflows to the German open-ended funds plummeted to just EUR3 billion, down from EUR13.7 billion the previous year.[23]

The path of the outflows marked out the direction of change. Investors quickly redirected their money away from poorly performing German-focused funds and towards those with greater international exposure and higher returns.

Not even the big tankers were spared. CGI lost EUR57 million of cash from investors in total in 2004 – a net effect of its flagship German HausInvest fund losing money while its global fund gained. Meanwhile, new players such as AXA and KanAm with no legacy portfolios in Germany calmly stayed in positive territory.

To survive, the funds needed to radically reduce their exposure to their homes market. However, the choice to unwind those positions at their own pace was taken off the table as they found themselves in a full-blown crisis. A sector with a 40-year reputation for not selling, was now in the unenviable position of needing to sell its worst-performing properties at the worst possible time.

## Making international inroads

Germany was a constant frustration to international investors. As the largest economy in Europe, it was a painful gap in any pan-European portfolio but with so little rise and fall in prices, it was a nightmare to find an entry point from whichever angle you approached it.

But with Germany now in recession and banks weighed down by anything from EUR200 billion to EUR350 billion of non-performing loans, there was suddenly a way in.

The potential to buy distressed assets was signalled by the presence of Lone Star, Cerberus and Fortress circling the country. These specialist buyers maintained a tight strategy focused on distress, trawling the world for opportunities, then closing up shop and moving on the moment they ran out.

Despite the distress, US investors already anticipated a rough ride. While other countries quickly capitulated under the pressure of opportunity fund money, the Germans were a particularly resistant strain of seller, determined to fend off this "plague of locusts" wanting to strip bare the financial sector.

Lone Star and JP Morgan bagged the first deal, buying EUR490 million of non-performing loans from Hypo at the end of 2003. The deal was more significant for the prestige of the seller rather than its size. If Hypo was willing to sell, others would follow. That said, it was not an easy bridgehead; pricing was in the range of 50 to 60 cents in the euro, much higher than the bargain basement 30 cent starting points in Paris[24] ten years previously.

A year later, the barriers lifted further. Lone Star paid EUR3.6 billion for a second loan portfolio from Hypo in the largest transaction of its kind in the world. There was no discount for size, with the bank negotiating a price "north of sixty cents in the euro".[25] Lone Star followed up with two portfolios from Dresdner Bank; a EUR1.2 billion loan portfolio and a EUR1.4 billion sale (scaled back from EUR4 billion) in joint venture with Merrill Lynch, while Morgan Stanley took another EUR400 million off Hypo's hands.[26]

Away from the non-performing loans sales, there were few other indications that Germany was ready to accommodate the kind of systematic buying campaign opportunity funds were after, even if there appeared to be increasing levels of distress. "Everybody thought from the outside 'There are great bargains to be had, we can be vultures and get everything for thirty cents in the dollar'," said one German fund manager. "Years and years these vultures circled around Germany getting so frustrated because the funds and the banks which backed them just kept the whole thing alive."

Even when the corporate sell-off trend landed in Germany, it only ended in a series of near misses. By early 2003, at least 12 German companies were cash-strapped enough to put portfolios on the block but only at book value, way too punchy for the opportunity fund buyers.[27]

As well as the high-profile failure of the Metro deal, the property arm of German technology group Siemens pulled a EUR400 million portfolio citing the disappointing offers.[28]

The only deal to make it through was the EUR1 billion sale and leaseback of Deutsche Bank branches by Blackstone, of which two-thirds were in Germany while the rest were in Spain, Belgium, Italy and Portugal.[29]

While the corporate buyers stalled, by 2004 it was more difficult for the public sector. At every level of government, politicians were tackling increasing budget deficits and high unemployment following years of economic stagnation. Selling off large chunks of property was within easy reach, and German Finance Minister Hans Eichel announced plans to raise EUR15.5 billion from state sales.[30]

Germany's public housing companies were first out of the blocks when Goldman Sachs' Whitehall fund and US opportunity fund Cerberus paid EUR1.97 billion for the GSW Group, a housing and construction company with 65,700 flats owned by the state of Berlin.[31] This deal was quickly followed by one almost twice the size as Fortress paid EUR3.5 billion for the Gagfah housing corporation, owner of 76,000 flats.[32]

MSREF also spent EUR2.1 billion on housing from industrial conglomerate ThyssenKrupp with German company Corpus Immobiliengruppe,[33] while Fortress added to its Gagfah portfolio with two further packages worth EUR3.25 billion.

This was the big investing theme that the opportunity funds were seeking to maintain their heavy-weight investment programmes across Europe. German housing offered billion-dollar deals in a restructuring play that passed large swaths of German property into international ownership. In the four years to 2007, more than 1 million apartments transferred into private hands.[34]

The scale of sell-off was sealed when Terra Firma bought the Viterra portfolio of 150,000 houses for EUR7 billion in Europe's largest ever deal. It was rumoured that the company's winning bid was EUR1 billion more than one of the rival bidders, and cemented Terra Firma's leading role in Germany's housing market.[35]

The deal added to Terra Firma's housing portfolio, which it managed through its Deutsche Annington company. It already owned a major residential portfolio in Germany, having bought the housing stock of state railway company Deutsche Bahn for EUR2.1 billion as early as 2001.

Further down the line, this group of determined buyers had ambitions to aggregate these deals into large German housing platforms. With the government now openly talking about the creation of a German REIT (G-REIT), the buyers were building up to exit their investment through the listed markets.

The raft of housing deals stunned the German public. Selling off state housing was highly political, cutting right through to the fundamental rights of the man and woman on the street.

Between them, the government and German industry owned 14% of the residential market,[36] and it was abhorrent that swaths of this housing would transfer into the hands of ruthless US vulture funds.

To mollify public concerns, the deals came with strings attached. US investors agreed not to carry out "luxury" refurbishments as an excuse to put up rents, and to restrictions as to when and how many units could be sold. Deutsche Annington could raise rents by no more than 3% a year and couldn't sell more than 20% of the portfolio within ten years.

Such restrictions made it difficult to see how they would generate 20% plus profits, and while the deals offered the scale the funds increasingly required, they were not going to be easy short-term wins. "These were very low-risk assets, most of these were large, boring portfolios with the returns mainly driven by the fantastic yield spread and not so much by the long-term prospects of the underlying assets," said Nick Jacobson, who worked at Citigroup, which advised on the Viterra deal. "You could generate very significant financial returns if you adopted high financing leverage or were able to capture significant operational synergies."

The buyers were looking to do both. These purchases naturally suited to the use of highly leveraged structures. The Viterra deal had already shown the depth of the traditional lending market for German banks including Eurohypo, RBS, Citigroup and Barclays Capital[37] providing around EUR4.5 billion of additional financing.

However, to move returns an average of 5% to 6% for German residential property (the initial yield for Viterra was 3.8%) to 20% or the high teens was a giant leap. Cerberus expected to make a 15% annual return from the GSW acquisition. "When you start laying on a lot of debt, you don't have much room for error. So, depending on exactly when you bought and exactly when your debt was due, you did well or poorly depending on that," said Peter Linneman.

To supplement returns, they counted on selling the houses to tenants, believing they could persuade the Germans to become owner occupiers in a nation of renters. It was a strategy that miscalculated the strong cultural attachment to renting, which with rent controls in place was as secure as buying a permanent home. "We thought a lot more Germans will want to own their home and we'll make an arbitrage between the rental value and the ownership value," said Peter Reilly, managing director at JP Morgan, which bought housing in Germany on a smaller scale. "It was easier said than done and there was a lot frictional costs of executing that strategy which ate into your profits. If you looked at the returns they were OK, probably in the lower quartile of all the deals we did."

## German funds struggle with distress

When DEGI failed to sell a EUR350 million portfolio for a second time in 2005, it became clear that the German open-ended fund managers were struggling to be distressed sellers. The price it was seeking was around EUR100 million above what any buyers were prepared to pay.[38]

They were not the only ones fooling themselves. A EUR300 million deal from DB Real Estate also stumbled as investors rejected three buildings in greater Frankfurt – or, at least at the suggested price – where the average vacancy was more than 18%.[39]

All through 2005, the pressure for the open-ended funds to offload poor-performing German properties was immense, and fund managers were making every effort to assemble appealing portfolios with one exception, the price. Constrained by their regulator, they were unable to sell properties for more than 5% below the book value. It was a paradox that saw fund managers admitting they were distressed sellers but still pricing stock as though it was the best days of the market.

Even with some fund managers modestly writing down portfolios, the book value was still likely to be overestimating any price validated by a market sale. The funds were just not traders. DIFA, the first to bring a German portfolio to the market, had sold only 133 buildings in the last four decades.[40]

Reluctantly, the financial parents of the funds were forced to step in and help. Within weeks after DEGI's EUR1.1 billion outflow, its parent company, Allianz, bought EUR1.8 billion of assets from the fund, which saved DEGI from having to put them on the market.[41]

They included the bloated Space Park project, which had resolutely failed to lift off. DEGI invested EUR500 million in the Bremen scheme only for it to close after six months, the German public failing to buy into the combination of space exploration and shopping.[42]

In another move, DB Real Estate and DIFA offered flexible portfolios, allowing potential buyers to cherry pick individual properties that could be sold off closer to book value.

Whatever they did, fund managers still found themselves locked into the straitjacket of their own valuation system and the tight regulation surrounding the funds. DEGI finally offloaded just two assets from the planned EUR350 million portfolio for EUR81 million,[43] while DB Real Estate only made more progress when it wrote down its buildings further by 4.6% to enable it to sell to Australian investor Rubicon.[44]

These were not the wholesale sales the German funds were looking for, and they were instead forced to turn to their collection of trophy assets across Europe.

Amid the rush of deals sprinting to close before the Christmas break in 2004, Deka sold the Lloyd's building. Its first purchase in the UK in 1996 for £180 million, now changed hands again for £231 million.

It was a deal that helped the UK market to an all-time investment record year of £35 billion,[45] but the significance of its sale as the funds' first significant international buy went unnoticed.

Other open-ended fund managers also reluctantly leaned on their trophy buildings for quick sales. CGI put One Curzon Street on the market, admitting that it was a "special asset that we know we will never replicate".[46] It sold to the Abu Dhabi royal family for £280 million, its yield of 4.1%[47] breaking new records for London's West End.

While these sales helped tackle the weakening returns and redemptions back home, they were merely chipping away at the edges of a growing problem. The outflows could not be stemmed as the German property market deteriorated and all efforts failed to restore the confidence of investors.

The response to DB Real Estate's news that it needed to devalue its property portfolio was so bad that EUR300 million drained from the fund in one day. It was forced to close the fund to investors, the first time one had been frozen in the sector's 40-year history.[48]

KanAm then received bad news from one of its US partners, The Mills Corporation, which announced that it needed to restate its financial position for the last five years. KanAm's association with the company caused a one-day EUR700 million run on its two funds, and it closed both to investors.[49]

In the two months straddling 2005 and 2006, investors pulled out EUR7.2 billion from the funds as the endemic nature of the crisis settled in. By now, 10% of the EUR85 billion industry was in frozen funds and many questioned whether the sector could survive.

The funds' regulator, BaFin (*Bundesanstalt für Finanzdienstleistungsaufsicht*, or the Federal Financial Supervisory Authority), started investigating reforms to ensure larger investors – mainly institutions – committed to at least one year, undermining the open-ended nature of the product. The sector's opaque style was also questioned, as it transpired that institutional investors were being given guaranteed returns over retail investors in some funds.

KanAm and DB Real Estate scrambled to sell assets outside Germany, an exercise now made less painful by the sheer heat of the market. KanAm sold five buildings in Paris to Morgan Stanley's French SIIC Mines de la Lucette for EUR1.1 billion.[50] Other bidders were on hand with more cash but Morgan Stanley made a EUR115 million down payment and completed the deal within a week. In Paris La Défense, DB Real Estate sold Tour Adria for EUR560 million, 37% above book price.[51]

This frenzy of international sales was at odds with its success in Germany, as they tried and failed to shift secondary buildings with higher vacancy and distress. But this was about the change as the force of international investing across Europe made Germany the unlikely last battleground before the crisis.

## The final battleground

The open-ended fund managers probably couldn't believe their luck as the mammoth wave of capital in Europe washed a second batch of foreign buyers into Germany. In late 2005, it was the UK investors that stepped off the plane.

In a complete role reversal of the first German arriving in London in the early 1990s, the Brits pitched up in Germany and started to judge the market on UK terms. There were no gleaming monoliths like the Lloyd's building to snap up but instead Lidl supermarkets or Metro DIY stores in secondary towns with 8% price tags hanging from their doors.

This was classic bottom-of-the-cycle stuff, or so they thought. After years of Germany being in the economic doldrums, Chancellor Angela Merkel was now in office promising much-needed economic reforms for the country. Europe's engine room was finally going to rev up.

This would help close the chasm the UK investors spotted between the higher-quality, well-let core investments favoured by the open-ended funds, and secondary retail properties. "The gap between those two markets was wider than almost anywhere else in Europe, which reflected the traditional bias among German institutions having very low costs of capital to focus primarily on prime office buildings," said Jeff Jacobson.

In the first half of 2006, international investors piled in. With European investment volumes already hitting EUR95 million, Germany suddenly took the largest share of cross-border deals, leaping over the UK and France. Overall, Germany was now second only to the UK in terms of total share of deals.[52]

The buyers were the private, more entrepreneurial-style players from the savvy Topland, which kicked off its planned EUR1 billion spree with EUR120 million of DIY stores,[53] through to lesser-known syndicators such as Liberty Land, which paid EUR190 million for a portfolio of properties across Germany.[54]

As successful as they were on their home turf, they were now pricing aggressively in a market where they had little understanding of the risks, and limited underwriting experience. With prices rising in the UK market, they were desperate to find yield elsewhere. Here, with yields close to 8% and borrowing costs at around 4%, they could continue their highly leveraged onslaught in cahoots with the equally voracious UK banks. "The red flag was that every single time we were doing portfolio acquisitions, whether German railways or the Hamburg portfolio, every single time we looked around the market for financing, it was RBS lending with very aggressive bids, very attractive financing and you thought where is this coming from?" said Richard Mully, who worked for fund manager Grove International Partners at the time.

The Irish banks were also hungry, backing their clients who never liked to miss out on an overheated market. Irish investors stoked the prime end of the market in Frankfurt to set record low yields for both retail at just over 4% with a Louis Vuitton store, and an office property at 4.7%.[55]

As the frantic grabs for stock continued, prices rose and yields fell as momentum kicked in. However, this was still Germany. Nothing fundamental had changed about this investment flatland except that tourists had arrived and they were making their own weather. "The power of overseas money moved the market in and the market itself never justified that low level of yield," said John Slade. "They created the cycle."

It was a cycle within a market with local investors knowing that the gap for those secondary properties may waver but they never fell because, in the end, they were priced correctly. There was zero demand for these types of property if the German retailers moved out, or even threatened to move out. But that was for the international investors to find out for themselves as they loaded up with leverage and carried on buying.

For the German open-ended funds, the market was moving up to meet them. With prices much closer to book value, they just threw EUR5 billion of property onto the fire and stood back.

Agents struggled to handle the number of portfolios suddenly coming to market as fund managers were desperate to capitalise on these hyped record prices.

While the funds had written down some of the values, this was going to be the first time they had the chance to sell some of their most difficult assets. One portfolio included "mega prime to stuff that makes you want to crawl into a corner."[56]

Deka soon became the first open-ended fund manager to sell a EUR1 billion plus portfolio, as Oaktree Capital took on 48 properties for EUR1.1 billion. One-fifth of the portfolio was vacant, and it also relieved Deka of EUR15 million in maintenance and asset management costs, of which only a portion could be recovered from the tenant.[57]

From then until early 2007, the open-ended funds sold another EUR6.1 billion of properties to US investors including Germany's second largest deal behind Karstadt, when Eurocastle, part of Fortress, paid EUR2.1 billion for nearly all of the domestic assets in DB Real Estate's troubled Grundbesitz fund. The price reflected a yield of

5.1%,[58] and the sale enabled DB Real Estate to turn its ailing fund into a European-focused fund, the preferred strategy from investors.

DEGI also more than halved its portfolio when it sold 37 assets worth EUR2.45 billion to Goldman Sachs' Whitehall fund, for a reported EUR2.6 billion, reflecting a yield of 4.6%.[59]

It wasn't all easy. The protracted nine months of negotiations for Deka's Hannibal portfolio saw it lose one potential buyer and Deka Chairman Reinhardt Gennies his job.[60] A EUR700 million version of the portfolio eventually sold to Dutch private investors.

For US investors, Germany was a macroeconomic play for growth but they ended up taking on debt-laden portfolios of secondary properties with huge levels of vacancy at the highest prices Germany had seen. "They leveraged the hell out of those deals, buying at 5.5%," said one private equity fund manager. "These deals looked good at a snapshot in time but there was no room to manoeuvre if things went wrong."

The sales transformed the open-ended funds, reviving a decaying sector and prompting the German public to begin reinvesting money in them. In the first nine months of 2007, they collected EUR6.9 billion of inflows and signed up for EUR8.3 billion of new purchases.

The funds were leaner and fitter, having shaken off large swaths of legacy assets that had been dragging down performance for years. With the patience demanded of them by the regulator and their valuation system, this was not done at too much of a cost. Those who thought that Germans didn't know how to do distress selling were right; this was a case study in just that. "I think they misunderstood the German resilience of Deutschland AG, the system," said one fund manager. "The whole thing was backed by banks and they would rather put in equity to save the sister than just give up and write things off."

# 9

# THE FINAL PARTY

MIPIM 2007 was a zoo. Officially, close to 30,000 property people caged in the small resort city of Cannes. Unofficially, thousands higher as unaccredited hangers-on joined the throng to hammer company expenses in the sunshine of the French Riviera. If the property market was white hot in 2007, MIPIM was just another place to burn money.

If anything said top of the market, it was all of this, everything happening in those four long days in March. Attendees complained about the hangers-on, could see the froth and feel the headiness, yet they also mocked the warning signs.

Every spare inch of the town was plastered with company logos, banners draped from the grand hotels and penthouse balconies along the Croisette beachfront boulevard. Corporate flags whipped above the parade of yachts bobbing up and down along almost half a mile of jetty.

Companies ate into seven-figure sums as they fought to make their mark among the commercial scrum. ING held cocktail parties each night as well as hosting clients on its yacht, in its sea-view apartment and at its stand in the exhibition hall. JLL and CBRE requisitioned beachside restaurants and erected elaborate hospitality tents. Airport lounge-style business centres for client meetings by day, by night pulsating white tents hosting late-night parties for the lucky few hundred.

In the evening, the late-night crowds descended from lavish dinners hosted by lawyers in the Old Town to the Hotel Martinez. The bar was six people deep as crowds clutched their EUR25 gin and tonics and EUR15 beers, spilling out onto the terrace, the hotel foyer and all the way to the other side of the street to Bar 73. For those who preferred their beer nearer and cheaper, the crowd was sprawled across the road at Bar Roma opposite the Palais exhibition hall.

The mood was resiliently buoyant, despite the unwieldy fair now fraying at the edges. The sheer volume of people saw attendees being bussed in daily from hotels as far away as Nice while spiralling flight costs meant the unlucky few routed themselves via Marseilles airport over two hours away.

Outside the exhibition hall, men on flip-phones looked around hopelessly, trying to locate the right other man on a flip-phone for their next meeting. The serendipity of casual encounters on the Croisette was replaced with the slalom of navigating the crowds. Female attendees staying resilient in the face of promotional women in catsuits handing out free mints.

The fair, usually dominated by the native French, as well as the Brits and Germans, was now overridden with money from all the exotic corners of the world. Staring out over Cannes from a 50-foot banner was a couple dressed in traditional Russian costume, calling for investors to sample the opportunities in the southern region of Krasnador Krai. A few hundred feet away, the region shared its specially erected marquee with Kazan, the self-styled third capital of Russia.

The presence from the Middle East was equally overwhelming: more hastily erected marquees to house first-time exhibitors with money flooding in from Bahrain, Kuwait, Qatar and Saudi Arabia. Dubai took pride of place, showcasing its latest project, the Burj Khalifa, that was rapidly ascending into the world's tallest building.

But in the end, nothing could break this crowd's self-congratulatory mood. Just two months earlier, the investment markets had wrapped on their biggest year yet, soaking up a 50% rise in volumes to EUR227 billion.[1] In effect, the European investment market had doubled since 2004 and now boasted bigger trading volumes than those in the US.[2]

There in the sunshine of Cannes, the industry preferred to be warmed by the success of the good year that had just passed rather than worry about any storms to come. Little did it realise that it was just weeks away from the deals that would define the top of the cycle. This was the last party.

## The run-up to MIPIM 2007

Life on the ground in 2006 was gladiatorial, a brutal fight for buildings where the only effective tactic was to pay more. Players faced off in bidding wars against 10, 20, even 30 rivals for single buildings and portfolios.

Those coming in second place blanched as winning bids equalled their own future exit scenario. "You know what a deal should price at and someone would sweep in and pay a 20% premium," said one fund manager. "You would be like 'have it' but you knew there was just lots of foolish money running around paying ridiculous prices in markets they had no experience in."

For those smart enough to be selling, they were finding the going vicious. "You put your hand out and they would chew your arm off," said Van Stults, one of the founders of Orion. "Selling was phenomenally good so we started pre-selling [developments]. We had two or three indicative deals where we sold at astronomical prices, empty buildings, just at completion with no rental guarantee at 50% above costs."

The gathering pace and velocity of the markets had wound the industry up into a cyclone of activity, sucking in everyone and dispatching them at top speed into distant parts of Europe. Investors trampled on each other to buy buildings in Central Europe, others ploughed straight on through, making big plays in tight markets: Romania, Slovakia and Croatia. The US opportunity funds anchored one foot in Germany and stamped the other in Spain, signalling their next and final major macroeconomic bet. Smaller companies clustered around them, snatching up anything in miniature that smelt just about the same.

Established players were tired of being burnt by the hot money of private investors and loose speculative consortiums. Rather than step back, they competed, paying over the odds in a one-upmanship of scorched-earth buying.

If Europe used to be a collection of domestic markets, then this was the new drill: no one, but no one, was at home. UK investors leapt boldly over neighbouring countries to debut internationally in Germany, Central Europe and beyond; the Spanish investors with their growing wealth set their sights on France and the UK. The Italians had plans as far east as Turkey. Pan-European fund managers carried the bags for Dutch and Nordic investors while the Irish spread themselves thinly right across the continent – anywhere Ryanair or their favourite banks could tempt them.

In just seven years, all the borders had fallen away, along with any inhibitions about capability or knowledge to invest on this scale. Everyone was trapped in the same behaviour; large or small, no one was exempt. The game was to max yourself out to the best of your ability in the pursuit of higher yield, even if that meant the locations got stranger, the quality of the buildings got poorer and the prices got higher.

"You can talk about capital markets and all those things, but one of the most powerful drivers of markets is momentum, pure momentum," said Richard Mully. "Why are you buying that asset at that price? Because it is going up and it's going to keep going up and other people are bidding it up. That's a pure momentum effect."

A consortium of Middle Eastern investors paid £100 million over the initial £315 million asking price for 33 Cavendish Square, in London's West End. They had recently lost out on another building in London and wanted to "avoid disappointment".[3]

In its first deal in Warsaw, UK fund manager London & Regional paid EUR260 million for the 40-storey Rondo 1 office tower. The unfinished building had already changed hands three times in just over a year, growing EUR64 million in price as it was handed from new owner to new owner.[4]

The further the industry galloped out, the more it caused the yields to fall. In 2004, in the major cities across Europe yields fell between 50 to 75 basis points. In 2005, they continued to narrow, falling a further 50 to 150 basis points. Coming into 2006, as the range of yields across different markets became strikingly similar, it was clear that prices being paid did not reflect the risks of the individual markets. The rapid expansion had emboldened players to ratchet up the risks. As the search for yield intensified, buyers spread to more emerging locations and settled on more secondary properties.

In Western markets, yields were positively squeaking. In Frankfurt, UBS Continental European Property and French company Shaftesbury bought a building in Mainzer Landstraße off a net initial yield of 4.5%. In London the Israeli-backed O&H Property paid £30.5 million for 12–13 Bruton Street in the West End for a 3.19% yield.[5]

In Central Europe, the size and frequency of deals was astonishing, particularly as the first ever sales to institutional buyers had taken place barely seven years before. Then, yields were 11%; now a Spanish investor paid EUR25 million for office building in the Polish capital at 5.4% benchmark, worryingly close to those of Western European office centres.

The German open-ended funds, which pioneered the Central European drive, were now frustrated and trapped – but not alone – in the furthest corners of Europe. "This is when I ended up in a southern Romanian city," said Thomas Beyerle. "And, we met all the other German [fund managers] there as well. There was just one flight a week."

DEGI lost out on that deal but was more successful with the Kings Cross Jankomir shopping centre in Croatia, paying EUR81.2 million off a keen 5.8% yield[6] for an immature market.

In Southern Europe, Madrid set a record yield of 3.5% – putting it in line with some deals in London – when private buyers paid EUR60 million for the Serrano 55 office building from Morgan Stanley. The yield was below that of Spanish ten-year bonds, with the buyer now assuming rental growth was on its way.

Climbing property returns were also drawing in piles of hot money, irritating new channels of cash attracted by the here-and-now of high returns, with no sense that they were already too late to the party.

Money poured in from private investors from Israel, Ireland, the Netherlands and Russia, often being funnelled through the UK's AIM, with 62 companies raising EUR9.6 billion in 2005 and 2006.[7] They all promised a gold rush from deals in the UK, later joining the throng in Germany, and in the new European Union ascendant countries. All the time they were mining fat fees off gross asset value on investments made with high levels of debt.

The monster feet of Irish private money were now trampling on the ambitions of even the largest investors as they swiped up chunky prime buildings right across Europe. In a display of the new-found individual wealth coming out of Ireland, a Dublin husband and wife bought 17 Columbus Courtyard for £120 million and 15 Westferry Circus for £135 million,[8] both deals part of Songbird's sales drive at Canary Wharf.

Irish investor Quinlan also made its Spanish debut buying Diagonal Mar shopping centre in Barcelona from Deka for EUR330 million, off a yield of 5%. The deal by Quinlan was agreed off-market, rather than being put to competitive bids, underscoring the strength of the offer.[9]

Funds became a factory line of products designed to soak up the unstoppable wave of interest from institutional investors. Capital raisers were still out on the road, but the phone at their desk was ringing constantly with investors begging to be let into the latest fund. "It was a rapidly moving thing," said Patrick Bushnell. "At the moment when it should have been impossible for us to raise money, was the moment it was easiest. We raised one fund in four or five months when its predecessor had taken us almost two years, and that was in 2005. Looking back, even with two years to run before the financial crash started, it was late in the cycle."

Accurately tracking fresh capital was impossible but it was clear that the equity was getting much harder to invest. In 2005, fund managers deployed 90% of the capital they raised. By 2007, that was down to 69%.[10]

By now, funds of funds had emerged, adding to the weight of money pouring into the sector. These were funds that pooled money from investors, and then took stakes in the range of funds being launched. They were growing fast, jumping from around 12 with EUR5 billion to invest in 2005 to 25 with EUR10 billion in 2006.[11]

These funds of funds enabled investors that were too small to qualify for the regular funds to club together to take stakes. It came at a cost with two layers of fees – one to pay the fund of funds manager and then fees for the underlying funds it was investing in – but it proved to be an efficient way for investors without in-house property teams to quickly gain access to the sector.

For fund managers, they were already a doubled-edged sword. Funds of funds often helped get funds over the line to close, but as they themselves reported to clients, they could be a tough customer. "Many fund of funds managers raised significant amounts of capital during the period and would invest in a first close of a fund and then keep topping up at subsequent closes as they had money pouring in at the fund of funds level," said Gabi Stein. "Then the pressure on the underlying managers from these managers to put the money out would be immense. They would ring up and say to managers, 'What have you bought? Are you buying anything else? We need this money to start working'."

Funds of funds were not the only ones piling on the pressure. Investors preferred their money not to touch the side of the fund, instead bouncing straight into deals. Speed to market produced heroes among the fund managers. "I still remember that in 2005 and 2006, investors said if the manager cannot deploy the money quickly, he's not successful. I would never say that again," said Michael Nielsen.

For fund managers, it was easy to grow but the pressure of running and growing their businesses was also setting in. It was expensive to maintain an expanding network of international offices across Europe, and the reality was that launching more funds just helped with this. "Whatever fund you set up, you would always get a fee. So you were always good to make money," said Pieter Hendrikse. Closed-end funds were easily the model of choice, trapping that fee for seven to ten years.

This cold reality was another reason for funds to clamber up the risk spectrum. "If you said, I'm prepared to invest with you across Europe but you need to make it an open-ended, zero-leveraged fund, most fund managers would have stopped and asked themselves: what is the opportunity cost of doing this? That cost might have been giving up the chance to create a far more profitable product with higher leverage. Perhaps that's why relatively few were created," said Neil Turner.

The fees for some funds were also based on gross asset value, hugely skewing the industry towards more leverage. This rewarded the fund manager disproportionately for market movement rather than property skills. By now, it was standard for value add and opportunity funds to be at 75% and 60% LTV for core.

But there were no complaints. With the availability of debt and the momentum from the markets, this was looking like a smart strategy. "In an upmarket everyone's a genius," said Jeff Jacobson. "At the end of the day in the early stages of a rising market all you have to do is be long and be leveraged to make returns. In many respects, the less you know about risk sometimes, the better, because the higher-risk strategies at that point tend to do better."

The industry understood that there was risk by using leverage in these types of deals; there just weren't the reference points for how that should work. "It just didn't have a way to frame it in terms of risk and return. I think people would do things but they didn't know if they were doing it core or core plus. They didn't distinguish. It's just broken, I fixed it and I made 100% return, is that good or bad?" said Ric Lewis.

That left fund managers shooting for the same returns for core and value added, and fighting against the momentum taking prices higher. "The risks you had to take early in the cycle to try to achieve that target return, versus the risks you had to take later in the cycle to try to achieve that same return were exponentially different," said Jacobson. "The typical investor will say, 'If I go into a value-added fund, I want 12–14% net IRR returns'. If values have gone up 30% from three years ago, it's not realistic to assume you are going to get the same return going forward or you are going to take on very different and greater risks seeking to achieve these same returns."

## Losing sight of the risks

By early 2007, the Bank of England was calling out the property industry as "potentially vulnerable".[12] Cut-throat competition was eating away at disciplined lending with LTVs still rising while interest cover, the ratio at which income can match interest payments, was falling fast.

Banks were driven by lending targets as a culture of "bonus-driven scoundrels" pervaded many banks with varying standards of lending across countries. "You would look at the name at the top [of the loan proposal], and there were places that knew what they were doing, places that just didn't know what they were doing, and places that would lie to you," said Simon Hersom, who worked at RBS at the time.

Either way, the culture fuelled riskier lending, with compensation for the volume of loans, not what you got back. "Of course, your best customer was the one you did most business with, and you did most business with the ones that you did most adventurous terms with," said Hersom.

The British and Irish banks were largely at the centre of the riskier lending but the Germans were under the same pressures. "I would say in 2006 and 2007, talking about the gold rush when everyone wanted to grow, some German banks lost some of their discipline," said Jürgen Fenk. "Eurohypo started to take mezzanine positions in the UK and US, and therefore went away from the lower-risk approach, and doing different things like 15-year loans on a seven- or ten-year lease maturity profile. That's where the stupid things have been done."

For the banks, the scrutiny as a public company was an added factor. "We had a lot of pressure driven by analysts, rating agents and investors on a quarter-by-quarter basis," said Fenk. "Banks like Hypo and Eurohypo, driven by capital markets, were almost forced to make mistakes, to go up the risk curve and do some deals because of higher pressure on profitability."

Harin Thaker, who worked at Hypo Real Estate at the time, said with prices shooting up and margins going down, some had questioned whether the bank should make a stand and stop lending but the leadership rejected the idea, knowing they would be pilloried by the capital markets. "It does look like the bonuses were driving it but I tell you capital markets put such pressure on delivering sustainable growth, scale and returns. That was actually what was pushing us."

This left banks just meeting unrealistic targets, unaware of how that added up as a whole for banks. "I'm not saying that as an industry that we have no responsibility but I can tell you that certain banks were desperate to lend against almost any asset in order to build their loan book. That was their job, first and foremost," said Richard Plummer, who founded Rockspring Property Investment Managers.

It left banks writing aggressive deals where they couldn't afford to get it wrong. "You almost have a fiduciary responsibility to take the best financing that you have even though you know the bank is selling you an option to walk away from your equity. You have all the upside, they have all the downside, which representing your clients looks like a pretty good trade. Sitting on the board of that bank you would be horrified," said Mully.

David Brush did a deal with Deutsche Bank in the Netherlands where JP Morgan had financed 95%. "It was crazy but when you think about it, it's a purely rational financial decision. It's not my job, and that's the problem with the financial industry generally, it's not my job to watch out for the other guy," said Brush. "I don't make money by keeping other people from being stupid."

Banks also had to contend with conduit lenders clipping at their heels as a dramatic increase in demand for CMBS bonds finally unlocked the potential of this specialist lending market in Europe.

As well as a growing league of buyers for institutional-grade CMBS paper – AAA through to BBB – from fixed-income fund managers, specialist asset-backed securities (ABS) buyers and banks, the market was seeing the rise of the collateralised debt obligations (CDOs). These CDOs bought up bonds across the ABS markets including CMBS, repackaged them and then used those bonds as collateral to issue a second set of paper based on the cashflow. These CDOs promised higher returns to investors – and created higher fees for the banks – as they packaged up the riskier tranches of ABS.

Even though this was already a pool of much riskier paper grouped together, ratings agencies tended to rated an AAA tranche based on its risk of default of what was now a hugely diverse pool of assets. It was high risk dressed up as lower risk, given credence through ratings.

The explosion of CDOs massively increased demand across the board for CMBS bonds, causing a sharp narrowing of AAA spreads – from 43 basis points to sub-20[13] – but also created a market for lower-grade paper, something which had never existed before on an organised level.

While none of this had anything to do with the property markets, it was soon a crank that was driving the cogs at the origination end of CMBS lending. Without a market for the junior tranches, the market had remained small and conservative. Now, with CDOs hungry for lower-grade bonds, originators could make bond issues work with any type of property.

The appetite for lower-rated bonds combined with tightening spreads was a culture in which conduit lending could thrive. Soon there were 21 conduits in the market, set up by investment banks, UK, Dutch and German banks, issuing EUR36.9 billion of paper across 51 issues in 2006, a steep rise from the EUR23.8 billion the year before.[14]

A few specialist property CDOs were also launched including by Eurohypo and Blackrock, buying up B loans directly from issuers, as well as BBB and BB CMBS tranches.[15]

CMBS remained just 10% of the lending volumes including the conduits but now it was disproportionately influential across the whole market. "The people now driving pricing down to irrational levels were the same people who couldn't get the product away five years ago, so were telling us banks we were mad charging [low] prices," said one banker.

Conduits became factories for tightly priced loans, particularly for higher LTVs. "I won most of my deals not on pricing. I won them on leverage. We were taking on more risks and we were getting paid for it. We were able to arbitrage that better than low risk," said one former investment banker.

Conduits took a high origination fee upfront and also made money between how much they charged for the loan and the cost to place the bonds in the market. Based on that model, they need a high turnover so became increasingly competitive.

Instead, some of the newer entrants to the conduit market were incentivised to churn through weaker stock purely for profit. "They were originating loans to be securitised and found they could sell B notes to CDO originators and make money that way as well. The banks said, now we know we don't have to hold any of that risk (i.e. the junior B notes), we can originate more aggressively because we have buyers for the B notes. That's when everyone started to get really sloppy," said Wilson Lee.

One conduit was making 3% on each deal. As they lent EUR100 million, they were making EUR3 million, and that money was recycled three or four times in a year. The Titan Europe programme from Credit Suisse as well as Talisman from ABN

Amro and Eclipse from Barclays drew most criticism for later defaults and how the loan and issues were structured.[16]

For traditional banks, competing with conduits was to pitch themselves again a completely different business model, and one whose success was itself being driven by forces from outside the property market.

Banks charged lower origination fees and depended much more on what was now a declining spread between what they were lending at and their funding sources. They were being driven to do less profitable business, pushing them into higher-risk areas. "There was a feedback loop," said Peter Denton. "Securitisation pushed mainstream banks into stuff they shouldn't have been doing such as extremely risky mezzanine lending."

The market became a sparring match between regular lending and conduits, with discipline losing out. Conduit lenders were prepared to lend to higher LTVs and able to take on more secondary property, but CMBS wanted a super-quick turnaround to lend then off-load the risk, leaving banks still in the running. "You would get beaten on loans by the more aggressive commercial banks all the time," said Nassar Hussain, who worked for Merrill Lynch at this time. "This was especially if there was a refinancing as opposed to an acquisition, as the borrower had more time to get a loan done with the commercial banks who tended to take much longer to execute a loan than an investment bank."

## The weight of capital

In the end, what the hype at MIPIM 2007 had really been about was an industry celebrating its place in the financial markets. The push forward, the momentum for growth of the last few years wasn't driven by demand, supply and rental prospects, but by weight of capital. This was an influx of money supported by low bond yields off the back of low interest rates.

Caught in between the shift from a domestic to an international property market, the industry aligned itself with this new benchmark, justifying its investment decisions based on where bond yields were rather than the micro-dynamics around its buildings. "Real estate does feel like it lives in its own little world with its eyes shut and with real estate as a business that was about just the bricks and mortar," said Rob Wilkinson. "I have to confess that if I look back to 2005 and 2006, [we were] not thinking enough about the implications of capital markets, and to some extent the political environment."

If the industry wanted to be judged by this benchmark then it really needed to understand why bond yields were so low. Far away from its bricks and mortar, it was now pegging itself to a European story, but one that was also part of a larger global narrative which would have major repercussions in the downturn.

The introduction of the euro enabled all eurozone countries to borrow at the same rate, sharply narrowing the difference in bond yields between strong economies such as Germany and weaker ones such as Spain and Greece. The European Central Bank (ECB) – and therefore commercial banks and other bond buyers – was treating the risk across the very different markets as equal. Commercial bank regulators were so confident in the halo effect of the euro, that they allowed European banks to buy bonds without setting aside equity for their purchases.

Soon, banks and other bond buyers – of which there were a growing number globally – were buying government bonds from the weaker countries to benefit from just a few more basis points spread on what now appeared to be the same risk. This helped sustain the low ten-year interest rates and encouraged the weaker economies to take on debt beyond the health of their underlying economies.

In search of greater yield, bond buyers were also snapping up innovative new products globally such as US sub-prime residential mortgage-backed securities (RMBS) or CMBS bond issuances.

Such was the grand scale – both politically and economically – of the euro, that its effect of levelling out the European terrain for bonds gave the property industry the same confidence to smooth out its inherent own risk of countries and local property markets as it continued to invest cross-border.

The industry had also worked hard to transform into a modern business: new products, the ability to track and benchmark performance, new levels of transparency across the board in the listed and private markets.

All these factors gave confidence to the growing volumes of capital, meaning that the risk premium that needed to be applied to property over bond yields narrowed[17] significantly. Shaping up had given it much more room for manoeuvre, and the capital markets had become a reason for continued investment without the usual security of the prospect of rental growth.

"Those [yield] spreads all narrowed because suddenly it's driven by weight of capital," said Joe Valente. "It's driven by global capital and that has all sorts of dangers. If you look around Europe, London will be driven primarily by global capital as everyone wants to be invested in London, it's a different type of market. So, rents will go up and down and they will have an impact on total value but by and large, it's that yield shift that is important." But this was not the same for all markets, such as smaller ones like Manchester, yet the mentality was the same. "The occupational market is what drives Manchester, not the weight of global capital, or even domestic capital."

The relative value of property to bonds saw a bounty of capital moving towards the industry. It collectively assumed that this was equity flowing into its new modern industry but with the low borrowing costs for banks and the ability to raise capital through CMBS – essentially offloading that risk, its wall of capital was actually debt.

"The biggest mistake collectively by the industry was that everybody was seeing endless liquidity but too few recognised it was actually simply easily available debt. The liquidity was in the form of debt and the prevailing notion was that you could eliminate risk concentration by securitising everything. Obviously, that's more likely true as long as you stay at 50% LTV. But what was happening was insane," said Michael Spies, senior managing director at Tishman Speyer.

The bigger you were, the greater the insanity. In the spring of 2007, intoxicated by hubris and driven by ego and bonuses, some of the biggest bets were placed at the highest point of the cycle. A potent mix of invincibility, competition and pressure to invest saw deals that resulted in some of the greatest destruction of value across the industry.

In the UK, Spanish listed company Metrovacesa bought the HSBC tower at Canary Wharf for £1.1 billion in far and away the UK's largest ever single-property purchase[18] for a yield of 3.8%. The sale and leaseback of the 42-storey building with the bank was funded by £280 million of equity and an £810 million bridge loan from HSBC.[19]

In Germany, Morgan Stanley completed its legendary Pegasus deal. Teaming up with German listed property company IVG, MSREF bought EUR2.56 billion of German property from open-ended fund Union Investments (formerly DIFA). MSREF took on 28 buildings for around EUR1.4 billion, and a separate property for EUR703 million, and IVG took 25 buildings worth around EUR495 million.[20] The sales price was 14.5% above the market value.[21] MSREF financed its share with 93% debt from RBS.

For some, Pegasus was a satisfying conclusion to an unspoken standoff between the German open-ended funds and US investors. "To have the lower quartile of your portfolio sold to people that think they can beat the German recovery is a fantastic deal," said one German fund manager.

Finally, over in Paris La Défense, the building that was the greatest casualty of the last crisis was now Europe's largest single property deal and on schedule to become a symbol for the global financial crisis.

Lehman was ripe for the classic investment bank quick-turnaround: a sliver of equity, a senior loan to be securitised, bridge equity, and advisory fees for arranging the loans and leading the securitisation. It shut down at least ten rival investors as it paid a premium for exclusivity to bid, cramming one month's worth of due diligence into one weekend to get the deal signed.

Lehman contributed just EUR71 million of equity with its local partner Atemi investing EUR6 million. The remaining EUR2 billion plus was made up by EUR475 million of bridge equity provided by Lehman and EUR1.64 billion of senior debt provided by Lehman Brothers' German division, half of which was syndicated to Goldman Sachs.[22]

First thing Monday morning, the rival bidders and the rest of the world found out that Lehman had agreed to buy Coeur Défense for EUR2.1 billion.

# 10

# CREDIT CRUNCH, AND DENIAL

The deal was in its final stages and the client was chipping. As a negotiating tactic, it was blunt and tiresome, the buyer chancing its arm that it could hustle a discount in the final hours before signing. The seller was expected to capitulate rather than see the deal fall out of bed.

This time, with Curzon Global Partners as the seller, they were going to let the tactic play out. Ric Lewis, chief executive, told his team to head back to the negotiating table and "sound really angry about [the chipping] but just make sure you let it happen."

Curzon sealed the deal to sell the EUR680 million portfolio of retail and offices on 7 August 2007.[1] It was a timely decision. Within 48 hours, the financial climate turned. French bank BNP Paribas froze the assets of two of its hedge funds that were heavily exposed to US sub-prime residential mortgages, leaving investors unable to withdraw money from the funds.[2] The credit crunch took its first bite in Europe.

What was at first a crisis about the failure of mortgages sold to low-income, overstretched homeowners in California, was now on Europe's doorstep. BNP had bought some of these bonds backed by sub-prime loans and now, without anyone to trade them with, they were imploding in its books as it was unable to give them a value.

It was the clearest sign yet that the confidence of Europe's banks to trade with each other had evaporated. A month later, the rate at which banks lent to each other was at its highest level since 1998, signalling that banks were worried whether the banks they lent to would survive, or that they themselves urgently needed money.[3]

It was this nervousness that Lewis and Simon Martin, then head of research at C, had feared. Back in early 2006, they realised that with the proliferation of securitisation and the rise of CDOs, they no longer had a good understanding of what was happening in the credit markets. Martin called a friend in the City, got invited to a lunch, met with some hedge fund managers, and was eventually handed some research papers that decried the end of securitised credit. "If this is true and this is what people are doing, this is going to be a shit-show," he said. "Somehow we needed to persuade the clients – without freaking them out – that the right thing to do was to stop deploying capital and start selling."

Lewis not only agreed but said he'd just come off a call with a fund manager that had wanted to buy Curzon's entire core-plus portfolio, and he was selling. At first, the buyer, which was wrapping the portfolio into a company being listed, was arranging 60% LTV debt for the deal. Then, when the IPO failed, the buyer opted for a different route, deciding to fund it instead with 95% debt. For Lewis and Martin, it was another sign. "If someone can go from 60% LTV to 95% in a heartbeat without a price check, we needed to sell everything," said Martin.

## Selective hearing

The sub-prime crisis of August 2007 was the early warning system. An alarm bell should have rung in the property industry's ears like tinnitus. This was global news. All day TV reporters looked startled in face-to-camera interviews as they tried to simplify and explain esoteric financial concepts that they barely understood themselves.

Like everyone else, the industry watched the contagion as bank after bank discovered toxic US sub-prime loan exposures in the back of their cupboards, and started revealing huge losses. The reach of the crisis started to feel unlimited yet, at this stage, very few people who worked in property questioned that it would have any effect on their own bricks and mortar.

The banks had bought US sub-prime mortgage loans as part of their investment in RMBS. Like CMBS, the collateral was property, but in this case backed by loans to homeowners. The US RMBS market was mammoth compared with CMBS but there was a major link in that the buyers for both were the same.

The CDOs gobbling up the B pieces of CMBS were also big buyers for RMBS, including those bonds issued off the back of sub-prime mortgages. At its height in 2006, the CDO market was US$1.1 trillion globally,[4] driving an insatiable need for bonds to buy, including CMBS and RMBS.

Sub-prime home loans were taken out by borrowers who were likely to default if there was just a small change in their financial circumstances. They were sold by confident and helpful salespeople, who – working on a commission basis – squeezed through as many loans as possible with no regard to the affordability for borrowers or how secure their credit rating was for lenders. That lack of interest continued as the mortgage companies rapidly packaged up the loans and released them into the bond market, as easily as letting go of a bunch of balloons.

It might have felt like several degrees of separation between sub-prime home loans and CMBS but the unfolding events were showing now that they were arm in arm, a relationship that had eluded even those working in them. "I operated in the European structured finance markets and because the European CMBS sector was quite specialist, I would have had little idea of the magnitude of sub-prime residential mortgage issuance in the USA," said Nassar Hussain.

## Different paths to the crash

That disconnect made for a haphazard 12 months as the industry blew through to September 2008. It was the oddest, patchiest of times, a full-blown crisis for some, others barely breaking their stride.

That difference in response exposed an industry that was siloed when it came to finance versus bricks and mortar. Those nearest the heat of the capital markets – the banks and the listed property companies – started to melt, whereas among the private buyers there was an inability or unwillingness to believe it might affect them.

CMBS was right by the fire. Within days the market was shut. It wasn't that bond yields had gone up to build in new levels of risk to pricing; there was just no appetite for the bonds at all. From being on track for a record year – already up 66% on the same period in 2006 at EUR37 billion – the market was now dead.[5] The casualties were high profile, leaving massive loans extended on record deals languishing on banks' balance sheets.

Lehman Brothers was forced to pull its EUR1.5 billion securitisation of the loan on Coeur Défense after it received very little interest. Instead, it raced to sell the bonds and reduce its exposure privately among its clients.[6] It also still had the bridge loan of EUR475 million as well as the small slice of equity all burning holes in its balance sheet.

HSBC was still saddled with the £800 million short-term loan to Metrovacesa for the sale of its own headquarters at Canary Wharf. Securitising the debt was off the table, and even the second option to syndicate was no longer attractive as margins on balance sheet loans had risen.[7]

For Lynn Gilbert, now at Barclays Capital, her view into the bond market through CMBS lending prompted her to move fast. "I had two reporting lines, one into the head of investment banking and one into the head of bond trading. We all looked at each other in August [20]07 and said, 'this is not going to get any better'. The markets are going to shut, we need to batten down the hatches."

The bank went into full syndication mode to sell off its EUR7 billion book, observed by Peter Denton who had joined Barclays just weeks before. "The reason I know that Barclays was early was because it syndicated so much to the German banks. The Germans did not stop buying until the third quarter of 2008," said Denton. "It was like a shadow year where there were a few thoughtful banks just selling all of their books. Morgan Stanley, us and then there were banks like CSFB and BAML, which – just because they were so pregnant – couldn't move. They certainly didn't react as quickly."

The sub-prime crisis was quick to expose the European banks' reliance on the securitisation markets. They were now a long way from the days when they initially eschewed CMBS in favour of building up their loan books.

Now, fully signed-up to securitisation, they were experiencing the impact of the financial turbulence that was that first test of these new debt products. The inevitable reality that just because risk had been shared around did not mean that there was no risk.

Banks immediately saw their profits being eaten away by the fallout of subprime. In the first six months, they suffered losses globally of EUR127 billion.[8] Hypo Real Estate reported a EUR390 million write-down in its 2007 results due to its CDO portfolio, followed by a further EUR175 million the next quarter,[9] while Eurohypo took a EUR203 million hit on its US portfolio[10] in the second half of 2008.

With the CMBS exit route blocked, this seriously curtailed the banks' pace of lending. It was an immediate jolt back to their former selves as they started to only make loans that they would be proud to keep on their balance sheets. Debt was back within the family. Banks became more discerning about the level of risk they were prepared to take on, only lending to their best clients, with few open to new business or clients without a pristine track record.

Deal sizes dropped as the comfort zone became EUR60 million to EUR120 million; any bigger, and old-style clubbing together of banks was the only way to make it work. Conversely, few markets were off limits and they were still open for business in more emerging locations such as Central Europe and Russia.

Greater levels of equity were needed to get deals done as LTVs dropped to a still-lofty 80%; in some markets such as the UK, Spain and Ireland, that was less than 60%.[11] Pricing also started to reflect risk better as margins moved out to 150–200 basis points, a long way from the "one basis point for each percentage LTV" in the previous couple of heady years.

Lending capacity was at an all-time low. Loan volumes in the first quarter of 2008 were less than one-tenth those a year before. Naturally, without easy credit, transaction volumes had also fallen, now standing at EUR37 billion for the first quarter of 2008 compared with EUR58 billion the year before.

But for borrowers, it wasn't just about securing new lending. For those carrying the EUR30 billion of loans due to mature in 2008, there was a struggle to refinance in this new regime. Those borrowers with limited access to top-up equity, or with large loans, which banks no longer wanted to hold (or could securitise), were looking particularly vulnerable.[12]

The list of banks not lending was generally longer than those that were, but many of the familiar names were still open for business: HBOS and other UK banks including Abbey and Lloyds TSB, as well as the Irish including Bank of Ireland, Allied Irish Bank and Anglo Irish Bank were still in the game.

RBS was continuing to lend voraciously, its appetite growing as it set out to prove that its megalomanic EUR70 billion joint takeover of Dutch bank ABN Amro with Belgian group Fortis and Spain's Banco Santander was worth the price paid, and the sharp hit to its share price.[13]

German banks such as Eurohypo, HSH Nordbank and Bayern LB also kept lending but reverted to the relative safety of senior loans that could be funded by *pfandbriefe*, their gilt-edged source of capital. "It didn't stop new business but I think it certainly helped take a bit of steam out of the end. But, by then, it was too late," said one banker.

Others called time on the market. In late 2007, Hypo was closed for new loans, saying that with its targets reached for the year, it would step back until 2008.[14] The news was a surprise considering its recent successful EUR5.6 billion bid for Depfa bank,[15] Europe's second largest provider of public sector finance.

Hypo's enlarged business was meant to offer safety in scale as well as access to new customers and a renewed capacity to write large loans across the continent. However, the strength of this new partnership was quickly undermined when in spring 2008, after profit falls, the bank sold a 24.9% stake to a consortium led by private equity company JC Flowers for EUR1.1 billion.[16] Other members of the consortium included Shinsei Bank and Grove International Partners, run by Richard Mully and former Goldman Sachs partner Richard Georgi.

Investment banks were in wash-up mode, scanning their books for existing loans to sell that they hadn't been able to offload before the sub-prime crisis hit. By now, conduit lending teams had been dismantled and no new business was being written. Morgan Stanley, Lehman Brothers, Bank of America and JP Morgan were among the sellers; Credit Suisse was marketing a large portfolio of loans at a discount of up to 15%.[17]

Bloated traditional banks were doing the same. HSH Nordbank sold EUR7.6 billion of loans in three credit default swap transactions. These were taken up by Hypo, BNP Paribas and Lehman, and saw the risk but not the loans transferred to new owners, with HSH receiving a fee for every quarter the swap was in place without defaults.[18]

The credit crunch also coincided with preparation for Basel II regulations, which came into force at the beginning of 2008. These revised the capital reserves to be kept back by banks to safeguard their lending, but inadvertently exacerbated the banks' risk positions by prompting a shuffling of senior and junior loans between banks.

Under Basel II, it was to become more expensive to lend on, and hold, junior loans. However, rather than shedding these from their books, syndication teams lightened their load by shifting what sold in the post-sub-prime market, which was less risky senior loans. "They were selling out of large amounts of senior debt, which was going to be low capital requirements. Therefore, the junior was being kept and was getting more and more toxic," said Simon Hersom.

Basel II also fed through to current lending. Bankers keen to keep lending at high LTVs reverse engineered what would be allowed under regulations, and then lent just below that.

With the senior loans no longer stretching past 60% LTV, new players soon moved into high-yielding mezzanine, bridging the growing gap between senior debt and equity, at a cost. Mezzanine lending was anywhere between 300 basis points and 1,000 basis points,[19] but for some looking to refinance but needing more equity, it was the only viable option.

Lehman launched a $1.5 billion mezzanine fund, and Apollo followed with a $1 billion fund both to provide mezzanine and to buy subordinated debt at a discount. There was an estimated overhang of up to EUR40 billion of loans, which banks had failed to securitise before the sub-prime crisis hit.

GE Real Estate (formerly GE Capital), meanwhile, was taking the opportunity one step further. It announced that after building up $22 billion in equity-led investments, it would now focus only on buying debt, starting with two loan portfolios including one of EUR1.3 billion. It said there was now a two-year window to buy non-securitised loan portfolios at a discount, taking the company full circle to its arrival in Paris in the mid-1990s.[20]

## Warnings from the listed market

If the seizure of the bond markets was not signal enough, the listed market had been waving its arms in warning since the beginning of 2007. Its prescient powers were legendary, usually foretelling the downturn in the property markets by about four to six months. Now, it was not signalling a looming credit crisis but rather a massive change in sentiment for the pricing of property.

Harm Meijer, who worked for JP Morgan at the time, attended a Christmas lunch in late 2006 where everyone was asked to predict what was to come for 2007. "The two words coming out of that meeting were 'capital flows'," he said. "Everyone had inflows into their funds and no one could see it turning because of that. December's share prices were very strong and then the UK market literally turned on 1 January 2007."

It was an expected reversal after a straight six-year bull run, rounded off with a 49% rise in share prices[21] in 2006. Every European country except one had out-performed the wider equities market.

The timing could not have been worse. After more than a decade of lobbying the UK government, the property industry was finally rewarded with the launch of its REIT on exactly the same day.

The UK REIT was part of a second wave of tax-transparent companies that was heralded as the coming of age for the sector (again). The European listed market was now EUR153 billion,[22] almost three times its size in 2003. Despite the growth, it was still seen as underdeveloped; European listed property accounted for 10% of the global sector whereas the direct market was 30%.

Now, it was hoped the arrival of REITs in the UK, Germany and Italy would follow the example of the French SIIC. After being introduced in 2003, SIICs dramatically increased the market's scale and liquidity, and France now boasted 42 companies with a market capitalisation of EUR40 billion.[23] REITs in the UK and Germany were predicted to add another $100 billion within five years.[24]

Another sign of the listed market's maturity was the emergence of focused dominant companies, led by Unibail, the original company behind Coeur Défense. Now, it was buying Rodamco Europe to become a EUR21 billion company, the largest by market value.[25] "Property companies started to become more a business instead of being a passive fund," said Meijer.

Against the backdrop of the bull run, private companies were also taking full advantage of a rare window to list. There were more than 25 IPOs in Europe in 2006 raising EUR4.75 billion, twice the previous year. Headlining this rush was the EUR650 million debut on Euronext of 30% of the non-listed fund Prologis European Properties, the largest IPO in Europe since Canary Wharf.[26]

US investor Fortress also sneaked a partial exit from its major portfolio buys in Germany, sidestepping the advent of G-REITs by selling 20% of its Gagfah housing company through a regular listing on the mid-cap MDax index.[27]

But the expected rush for US investors to list their housing portfolios in Germany was quickly shot down by the government when it excluded resi-. dential companies from G-REITs and restricted those listing to a conservative 55% gearing.

It was Alstria, which was founded on property bought from the City of Hamburg, that became the first G-REIT after it converted from its regular listed status in October 2007. "The genesis of that was a private equity deal unlocked by American capital which saw the value of owning this [property] and then a bit of capital markets arbitrage taking it from private to the public markets," said Mully, who was one of the original buyers with Grove.

But the market was no pushover. Lehman failed to get the price it wanted for Uni-Invest in 2007, a Dutch listed company it took private in 2003, which was now stripped back to only the secondary assets. Meanwhile, Germany's largest existing listed company, IVG, halted its plans to convert to a G-REIT in late 2007, preferring to wait until the market improved.[28]

As the credit crunch added to investors' concerns, a black hole opened up in the listed property sector in Spain.

There was something about the Spanish companies' particular brand of largesse with their domestic and then pan-European ambitions that did not combine well its humble roots of house building and land banking. Property stocks fell quickly as investors realised housing in Spain was hugely overvalued. Spanish banks had adopted similar sub-prime lending trends, extending loans to those on a low income or with a bad credit rating. While this only made up just 1% of all mortgages, the knife-edge lending was a rippling concern through the sector.

The economic boom brought on by the euro saw Spain experience some of the fastest growth in house prices and lending in Europe.[29] The pace and scale of this had driven on the ambitions of the major housebuilders, but now with the value of the land banks which underpinned their wealth being wiped out, it exposed the old-school leadership. "They were all run by big ego guys, old Spanish gentlemen building their empires. They had been making massive money for a long time, trading land, building homes and they had had it very easy for a long time. They thought they could turn everything into gold," said Jeppe de Boer.

At Colonial, the Spanish regulator suspended its shares after a freefall of 46% in a matter of weeks, as major shareholders dumped their stock. It sent the company into full-scale fire-sale mode trying to pay down some of its EUR8.9 billion debts, equivalent to two-thirds of its NAV of EUR13.6 billion. On the sales block was Colonial's 24% stake in French company SFL and its 15% stake in Spanish company FCC.[30]

Metrovacesa was embroiled in a two-year power struggle worthy of a soap opera, culminating in a convoluted and cumbersome share swap to split the company between the two major shareholders. After a 63% share rise over the previous six months, the company was now down 17%, wiping EUR1.1 billion from its EUR12.6 billion market capitalisation.[31]

Its impressive spending spree was running out of steam despite its bid to be the largest property company in Europe by 2010.[32] It pulled out of buying a EUR2 billion Germany office portfolio from Allianz, which was later snapped up by Goldman Sachs' Whitehall fund.[33] It also deflated a bidding war by stepping back from the battle for a EUR1.2 billion Corio portfolio in the Netherlands, leaving Lehman as the only buyer standing.[34] Both these withdrawals were taken as signs by the market that the Spanish company was just buying for as long as it could find banks to finance it.

By the spring of 2008, Metrovacesa was close to putting the HSBC tower at Canary Wharf back on the market, after failing to refinance an £810 million short-term loan from HSBC. In just a year, the value of the building had fallen to £800 million, leaving it with an LTV of more than 100%.[35]

## Lessons in denial

Whatever was being seen in the financial markets, many buyers were still partying like it was 2006.

In 2007 – before and after the sub-prime crisis hit – the Irish were the second biggest investors in Continental Europe, splashing out EUR13.9 billion on property across the UK, France and Germany.[36] Irish investors appeared to side-step the credit crunch, feeling immune with Dublin's strong financial sector still unaffected by the shifts in the global markets. As they were mostly investing personal wealth, they also tended to assume themselves to be insulated from the liquidity squeeze.

These were often colourful characters with rags-to-riches stories, whose trading nous had amassed them an astonishing personal fortune. Derek Quinlan worked as a tax inspector before offering rich clients financial advice, which morphed into becoming a dealmaker for the Irish elite, including buying the Savoy hotel group for EUR1.1 billion in 2004.

He was now fronting Herculean buys. He teamed up with PropInvest, led by lawyer-turned-property tycoon Glen Maud, to buy Citigroup tower at Canary Wharf for £1 billion, the second largest single-asset deal after the HSBC tower sale to Metrovacesa. The joint venture was not even rattled as completion was delayed by the sub-prime crisis. It closed in late 2007 with £875 million of senior debt from a consortium led by Allied Irish Bank and Banco Santander,[37] and a £204 million junior loan from RBS, also the seller of the building. By the time the deal was finalised, the building was valued at £1.2 billion,[38] the high level of debt showing that LTVs of 75% and above were still possible.

In Paris, Irish investor Sloane Capital paid out EUR650 million for 9 Place Vendôme in Paris,[39] a mixed retail and office building, setting a record low yield of 3.5% for the French capital.

By now, the Irish were outbidding other major high-leveraged buyers. In Spain, PropInvest took out Goldman Sachs, JP Morgan and Metrovacesa to pay EUR1.9 billion for Banco Santander's Financial City headquarters in Madrid. The deal, thought to be the largest ever in Spain, made the bank a EUR605 million profit.[40]

The Germans too had re-found their form. SEB had been selling its assets in Spain and the UK from 2006 onwards, leaving the fund sitting on a pile of cash. Now with equity back in vogue, it set its sights on the heart of one of Europe's most exciting cities. The development of Potsdamer Platz, a square that had been divided by the Berlin Wall for close to 30 years, was seen as the physical reuniting of East and West, spending much of the late 1990s as Europe's largest building site. Now, SEB had bought a complex of 19 buildings including the Grand Hyatt Berlin and the Potsdamer Platz Arkaden shopping centre, for EUR1.4 billion[41] in a renewed show of might.

The only place visibly slowing down was the UK, which was usually the place to see the end of the cycle first. Already in August 2007, the market had seen a 0% return, the first time it hadn't been positive since late 1992.[42]

Smaller private investors were spooked first. They had been pouring money into retail funds managed by companies that also invested institutional capital. They had also been lured into investing with some young pretenders that had come into the sector hard and fast.

New Star Asset Management had become a household name, luring in the UK public with promises of high returns in daily traded funds on billboards around the country. Set up in 1999 by John Duffield, an aggressive and successful financier who made £185 million from Jupiter Fund Management to Commerzbank. New Star ballooned to managing £24.9 billion of money under his management in just six years.

As seeds of doubt were planted by commentators in the personal finance pages of the national newspapers, New Star became a bellwether for the sector. Within months its shares lost three-quarters of their value,[43] and an almost 18% drop in the value of the assets in its fund. This prompted a sharp reversal, with outflows of £500 million in the second half of 2007, against an inflow of £2.3 billion in the first half.

New Star was not alone as other funds were frozen after high outflows. Scottish Equitable told 129,000 retail investors in its £2 billion property fund that they could not sell for at least 12 months, and Friends Provident barred 180,000 policyholders from leaving its £1.2 billion property funds.[44]

While these retail funds were the frontline for the impact of the credit crunch, these were daily priced funds that needed to revalue each month, so there was a duty to respond fast in light of falling values.

The exodus, while bad, was partly caused by sentiment, destabilising that part of the market, raising the issue once again as to whether these sorts of funds were suitable for illiquid assets such as property.

The decision for the retail funds to quickly take a realistic view was getting less purchase in the institutional market. Instead, it was priding itself on its longer-term outlook, giving it cause to be less worried about what could be a short-term blip. Even though, during 2007, UK property values fell by 7.7%, they figured that as the credit crunch hit, the usual summer slowdown would cover the waters and there could be a return to normality in September.

Part of the reason no one in the UK wanted to believe the end of the cycle had come was because they preferred the good news that was emerging. The irony was that as the capital markets were stalling, property fundamentals were on their way back.

If you could drown out the noise around the credit crunch, the big news for 2007 was that rental growth was back in London and beyond. Rents were rising, or about to, in 21 out of the 24 European cities, with only three – St Petersburg, Kiev and Moscow – having seen rental growth slow.[45]

The amount of space being leased across the continent was also motoring. In the first half of 2007, it hit a record high of 6.7 million sq m, led by Paris letting around 1.4 million sq m of space.

Cranes had also appeared on the skylines again. Paris and Moscow had twice as much development than they did in 2005, while Berlin, Budapest, London and Madrid had also seen a strong uptick.

Unibail unveiled the first drawings of its 71-storey Tour Phare, a new neighbour for Coeur Défense, and in Moscow, the Federation Tower was set to be the tallest in Europe. In London, work had just started on The Shard at London Bridge, the UK's tallest tower.

Agents reported that it might not be five people to each deal these days, but it was still two or three. By the end of 2007, this was still a market that believed its buildings were half full, rather than half empty. "Many people jumped in thinking this was the start of the development cycle, and they had four or five years to play on this," said Martin. "But it turned into the mother of all credit cycles and [the industry was] mistaking rising prices associated with credit for rising prices in anticipation of rental growth."

## Countdown to Lehman

There were mixed messages from US investors; they could see the maelstrom in the capital markets from their desks but continued to be at the height of their investing power.

Morgan Stanley was now investing MSREF VI International, which had raised a record $8 billion in equity. With debt, it had buying power globally of $30 billion.[46] This record-breaking fund was coming just 14 months after closing MSREV V, which had invested $16 billion[47] around the world. There were already rumours of number VII.

Inside Morgan Stanley, there were also changes that ratcheted up the pace of investing to a new level. The funds business was separated from the advisory side and moved to sit with the asset management division.

While this boosted the bank's financial status, with Wall Street valuing the asset management division more highly, it gave the funds tunnel vision. Now, rather than being part of a wider property business, AUM was the only yardstick. The solution was "if it moves, buy it".

Morgan Stanley stayed focused on Germany where it had now sunk more than EUR10 billion.[48] It joined SEB at Potsdamer Platz buying the Sony Center, a retail and entertainment complex housed under a conical glass roof in the shape of Mount Fuji in Japan. It paid EUR600 million, a post-credit crunch bargain, down from the original EUR800 million asking price as part of its Japanese owner's financial restructuring. Its core fund also teamed up with its open-ended fund to buy the Trianon tower in Frankfurt for around EU620 million with 90% debt, reflecting a net initial yield of 3.8–4%.[49]

Goldman Sachs was also still active in Germany. It fought off competition from Deutsche Annington to pay EUR3.4 billion, of which around EUR700 million was equity,[50] for the LEG portfolio of 93,000 apartments in Germany. It was the country's biggest private equity deal for 2008 to date.[51] With residential prices on previous deals already down in Germany by 30–40%, analysts at Merrill Lynch said the sector was now overvalued.

Goldman also finally sealed the Charlotte portfolio for EUR1.7 billion, buying 190 buildings in Germany from insurer Allianz. The portfolio was scaled back to help secure financing from banks including RBS and HVB, which were now reluctant to finance any challenging properties.[52]

However, Goldman was either looking for a quick turnaround in Germany or having a serious case of buyer's remorse. It put its 51% stake in Karstadt up for sale, just over a year after investing EUR120 million of equity into the highly leveraged department store deal.[53] It was looking to exit along with Arcandor, the holding company of Karstadt and its partner in the deal. But with the credit crunch, the value of the portfolio had fallen, leaving Goldman unable to exit its mezzanine debt. Arcandor sold its 49% to a consortium led by Pirelli, RREEF and Italian retailer Borletti.[54]

Goldman also tried and failed to sell up to EUR1.2 billion of the EUR2.6 billion portfolio it had bought from DEGI just two months previously. This turnaround was so quick, the original deal wasn't even closed yet.[55]

As 2008 moved towards the summer, all attention focused on Lehman as with falling profits and a downgraded credit rating, there started to be concern about the health of the bank.

As Lehman struggled to right itself, the extent of its commercial property activities in Europe and the US emerged as integral to its precarious financial position.

Historically, property had been profitable for Lehman, so from 2006 onwards it took a strategic decision to commit more of its own money to the sector. In just one year, it almost doubled its on-balance sheet property assets from $28.9 billion in 2006 to $55.2 billion in 2007.[56] "Lehman was a debt machine, which was ultimately its problem," said one fund manager. But the bank was not stopping as bond buyers bought for yield rather than activity in the underlying property market. "Clearly, every time they did a new issue it was over-subscribed, but for real estate fundamentalists, that was just nuts," said another fund manager.

In Europe, Lehman's portfolio was worth $12.5 billion but its record-breaking EUR2.1 billion purchase of Coeur Défense barely registered against some of its activities worldwide, including its $22.2 billion acquisition of US residential REIT Archstone with Tishman Speyer in October 2007.[57]

Now, the bank's inability to sell on the debt had left it with $19.6 billion of bridge from just its top ten deals.[58] This included $2.9 billion for Coeur Défense and $5.4 billion for Archstone. Suffering a slow death as confidence in its survival faded, Lehman set to do what it could to reduce its balance sheet in Europe.

Its solution epitomised the complexity of property in this new age as it hastily packaged up its riskiest debt positions – those it had so far failed to securitise or sell – into a CDO called Excalibur.

Excalibur was loaded up with EUR2.16 billion spread across 62 different debt positions: senior and mezzanine loans, development loans, B notes, CMBS bonds, corporate bonds to real estate companies, as well as syndicated positions bought from other banks.[59] The CDO was shoved into the ECB's emergency re-purchasing "eurosystem", to raise some much-needed liquidity. The ECB passed it on to the German Bundesbank, which with the financial crisis descending had no choice but to keep the bonds on its books until appetite from investors returned.

At a corporate level, Lehman was now quietly talking to buyers for its global property portfolio valued at around $40 billion.[60] BlackRock, Blackstone and Colony Capital were among interested parties, but day to day all they saw was falling property values not being reflected in the price.

When this sale failed, its last-ditch plan was to spin off around $32 billion of its bad property assets into a new company – its own version of the RTC from France – as the US government and the Securities and Exchange Commission conducted high-level talks to sell the whole bank.

In the end, talks failed and on 15 September 2008, Lehman Brothers declared bankruptcy, engulfing the world in the global financial crisis.

# 11

# LIFE AFTER LEHMAN

Expo Real 2008 was a funeral. It was early October, barely three weeks since Lehman had died, and just the evening before the show opened German Chancellor Angela Merkel had addressed the nation on the unfolding financial crisis caused by the bank's crash.

Mourners at the annual property exhibition clustered in small groups amid the sparse concrete courtyards slotted in between the cavernous exhibition halls. Arms crossed and heads down, they swapped war stories of the crash so far.

The mood was sombre, but then this was Expo Real. Never the jolliest of jollies. Tucked away at the purpose-built congress centre at the old Munich airport, it was a far cry from the sea and sunshine of its troublesome French cousin. It was Expo Real for business, MIPIM for pleasure. Exhibitors stuck to their stands to receive visitors in half-hour time slots and lunch was beer and white sausages at a functional cafeteria at the side of the hall.

But in the last decade, Expo Real had carved its time into the industry diary on the back of the undeniable dominance of its two major property exports: the open-ended funds and the banks. Now, in the next three days both had to begin their battles for survival.

## Bailing out the banks

As the delegates pushed through the turnstiles on the first day of the show, it was to the news that a EUR50 billion rescue had finally been agreed for Hypo Real Estate. It followed days of wrangling – and stern interventions by Chancellor Merkel – by fellow banks and the state to agree to the bailout. It was the second plan within a week, the first failing when Hypo angered fellow banks by underestimating its problems by EUR15 billion.[1]

Even then, there were fears that the EUR50 billion could still just be an expensive sticking plaster with the bank's long-term future only expected to become clear in the next few days. The share price continued its precipitous fall. When the JC Flowers consortium had invested EUR1.1 billion six months previously, it was at EUR22.50 a share; the price was now EUR4.82,[2] reducing the bank's market value to less than EUR1 billion.[3]

Over at the Hypo stand in the exhibition hall, there was a sense of practised calm as people gathered and chatted quietly on its elaborate multi-level staging. Then at its next door neighbour, there was a similar underlying tension as the exhibitor, RBS, also awaited its fate.

That day, one of RBS' main corporate clients had withdrawn a multi-billion-pound sum so substantial that it shook the markets and brought the bank to the brink of insolvency.[4] Its share price almost halved to 85p over the next three weeks as the UK government scrambled to secure a rescue plan. The next month, it took a 58% stake in RBS for £15 billion.[5]

Later, the bank announced for 2008 the biggest corporate loss in British history, totalling £24.1 billion, and that UK taxpayers were to contribute an additional £13 billion of cash.

With two of the property industry's leading lenders in desperate financial trouble, the depth of the entanglements the property industry had in the unfolding financial crisis was coming into focus. Not only would the ramifications of poor and excessive lending filter down to borrowers, but the sheer extent of that lending had made the industry complicit in the banking sector's weakness.

For Hypo, its undoing was less about the day-to-day lending and more about its purchase of Depfa bank in 2007 for EUR5.6 billion. Depfa was running a classic mismatch, financing its long-term loans for infrastructure and the public sector with short-term money from the wholesale market. Following Lehman's collapse, its sources of short-term liquidity just evaporated. "The day I joined the board of Hypo Real Estate, it was mid-August 2008. Hypo Real Estate had cash, liquid assets and credit lines available of EUR50 billion," said Richard Mully. "Nine weeks later, they had no working capital lines and credit lines of zero."

The equity of around half of the EUR1.1 billion the JC Flowers consortium had invested in the bank was immediately wiped out, which had been a shock to the consortium. "The investment had been made at a substantial discount to our estimate of book value based on a liquidation scenario at stressed prices," said Mully. "This was derived from a detailed line-by-line review of the value of the real estate lending book, which we believed to be pretty solid, and capable of withstanding a severe downside valuation case. What we underestimated was the liability side of the balance sheet.".

By April 2009, after a total of EUR100 billion of state aid, the bank was nationalised by the German government, giving shareholders just EUR1.39 share, and squeezing the JC Flowers consortium out of the deal. Hypo later changed its name to Deutsche Pfandbriefbank and hived off EUR210 billion of bad loans to a separate bank.[6]

The fallout for high-profile banks continued across Europe as the systemic impact travelled through the major banking systems in Germany, Ireland and Spain.

In Germany, Eurohypo's troubles arrived more slowly as its exposures to sub-prime residential continued to catch up with it. At the end of 2008, it reported a EUR1.4 billion loss for the year, compared to a profit of EUR588 million the previous year.[7] It would take a more organic approach to mend its business, shrinking its loan book by 20% over five years, reducing new business and refinancing loans.[8] Three months later it retreated from 20 countries, sticking to countries and business where it could generate returns over the long term and with what it saw as an adequate level of risk.

In Ireland, the largesse of the banks' lending to property both at home and abroad soon sideswiped the Irish economy as the drying up of liquidity threatened the viability of the entire banking sector.

The Irish banks had turned away from the traditional route of funding loans from customer deposits. Instead, they took advantage of their place in the EU to access cheap funding, becoming reliant on the wholesale markets to borrow short term and lend long.

As those sources of funding disappeared, confidence in banks plummeted, added to by the Irish economy being the first in Europe to go into recession.

The Irish government stepped in as Anglo Irish's share price dropped from EUR17 to 22 cents in just over a year, and, fearing it would collapse, took the bank into national ownership.[9] By the end of 2010, the government owned 93% of Anglo Irish as well as stakes in Allied Irish Bank and Bank of Ireland, overall injecting EUR64 billion of bailout money[10] into the banking sector.

Property was at the eye of the storm, as banks' loan books breezed pass regulated limits on property unchallenged. They had lent aggressively on development at home as well as heading into foreign markets with inexperienced borrowers, offering them 100% LTVs.[11] In late 2009, the Irish government established the National Asset Management Agency (NAMA) into which it would transfer EUR74 billion of mostly non-performing loans. This lightened the banks' balance sheets, at a cost to the Irish government of around 45% of the loan values.[12]

In Spain, regulation prevented the domestic banks from investing in US sub-prime loans, saving them from that contagion. However, their troubles lay closer to home as house prices fells amid economic worries about oversupply. Soon, Spanish banks were buckling under the strain of the EUR300 billion of loans to property developers.[13] The market had 920,000 unsold homes,[14] and to survive, the banks found themselves agreeing to refinance debts or buying unsellable assets from the companies.

It was a vicious cycle. If the developers collapsed, this increased the banks' debt levels and lowered their investment ratings, making it more difficult for them to continue to secure financing.

Alongside the banks' measures, the Spanish government created funds of up to EUR50 billion to shore up the banking system – although it would only take on healthy assets and not toxic ones. It also took steps to allow it to buy up stakes in banks to strengthen their equity.[15] However, it was reluctant to step in, instead asking banks to extend loan periods or stall interest payments and kick the can down the road.

Spanish banks bought developments from domestic players, swapping debt for equity and creating new property companies in a repeat of the last 1990s downturn. Banco Santander created Altamira Real Estate while BBVA set up Anida to manage the properties it was taking on. Later, more than half of Spain's 42 regional savings banks joined together to create Ahorro Corporación Soluciones Inmobiliarias (ACSI) to offload EUR3 billion of assets, to sell defaulted property, land and non-performing loans at discounts of up to 20%.[16]

The Spanish government followed NAMA's lead in 2012 when it set up a "bad bank", the Sociedad de Gestión de Activos Procedentes de la Reestructuración Bancaria (SAREB), and transferred just over EUR50 billion of properties from the country's bailed-out banks.[17]

## Closing in on the open-ended funds

Back in Germany, alongside her finance minister, Peer Steinbrück, Chancellor Merkel reassured the German public that all their savings accounts would be guaranteed by the government. "I was sitting at home in front of my TV, saying 'what does it mean'?" said Michael Birnbaum. "The next day, we ran out of paper because so many faxes came in with people terminating their investments."

In her bid to allay people's fears and the financial crisis, Chancellor Merkel had just inadvertently wreaked havoc on the EUR89 billion German open-ended funds industry. "What they really feared was a run on banks and that was the way to calm that down," said Barbara Knoflach. "Everything else was not really taken into consideration."

The omission was costly. That Monday, KanAm lost EUR400 million of investments from across its funds. Three weeks later, it closed its EUR580 million US fund and then its EUR4.9 billion European fund. Within days, SEB suspended redemptions on its funds, quickly followed by DEGI, AXA, UBS, and Morgan Stanley.[18]

In the year in the run-up to the collapse of Lehman Brothers, it felt like order had been restored for the world of low-leverage equity investing by the German open-ended funds. With more than EUR22 billion[19] to spend, they had once again become the saviours of the markets. In the year to July 2008, the funds bought 271 new assets for EUR14.4 billion, up from EUR8 billion.[20]

In London, it could have been the early 1990s all over again. Cash-rich and reinvigorated after a clearing out of their secondary buildings to international investors back home, the German open-ended funds were on the hunt again. By the end of the summer, they had already spent £1 billion in the City of London alone. With values in the UK market declining since the sub-prime crisis, prime yields were back to around 5.75% and there were willing sellers as cash-strapped UK funds looked to get some cash back in the bank.

Deka bought Moor House at London Wall for £230 million, 50 Finsbury Square for £113 million and 1 Old Jewry for £85 million within a matter of months,[21] while SEB agreed to pay £180 million for 88 Wood Street.[22]

Their buying was also not restricted to the UK. KanAm was preparing to buy the 42-storey OpernTurm from Tishman Speyer for EUR550 million, its debut in its home market.[23]

Now, within less than a month after Expo Real, 12 funds holding EUR34 billion – more than one-third of the sector's assets – were frozen to investors for three months. Eight of them later extended that period for a further nine months at least. In October 2008 – the same month as Expo Real – investors withdrew EUR5.1 billion, wiping out all the inflows they had pulled in since the beginning of the year.[24] It was happening so fast, that the funds were close to breaching their 5% minimum liquidity buffer.

The OpernTurm sale was off, as was SEB's purchase of 88 Wood Street and German open-ended fund managers joined bankers in the misery of Expo Real. "That was a bit spooky," said Knoflach. "Usually you have all these big parties. The people were all there but there were no parties."

## Older funds win out

While the general public acted fast to transfer its investments to protected savings banks, it was institutional investors and funds of funds that created havoc by removing huge chunks of cash immediately. Some of these investors had been using the funds as a cash equivalent, others were now overweight in property and needed to sell their units to rebalance their portfolios. "They would withdraw the money as soon as they could," said Knoflach. "This type of fund of funds investing heavily in the market just created a much higher turnover in and out. The wave in and out was much higher than everyone was used to before."

Some larger open-ended funds were able to protect themselves as they had created special funds for institutional investors or share classes which had redemption notices and exit penalties, but for most the outflows from their investor base snowballed.

Just three fund managers kept their funds open to investors – Union, Deka and CGI. Even as the largest and most established, they admitted to buying back their own shares to shore up their available capital. However, longevity and direct contact with a loyal investor base was now winning over the fleet-of-foot newer funds. The old guard soon revived with returning inflows, but for the rest, it was becoming clear that the blow may well have been fatal. By the middle of 2009, one of the newer funds, Morgan Stanley's P2, was fading fast after a 10% write-down in its value. Globally, its investments were suffering: 20% down in the US and Spain, 13% in Japan and 12% in Germany. In Madrid, it wrote down one property from EUR121 million to EUR89 million. The joint ownership of the Trianon building in Frankfurt with its core fund cousin was also faltering. Its value had fallen 13% from EUR620 million but its costs had also risen 15%.[25]

Some funds dipped their toes into the market again in July only to experience further outflows. Credit Suisse's Euroreal lost another EUR778 million that month,[26] while at KanAm, another EUR421 million drained out.

The funds were reliving the nightmare of just three years before when corruption and poor performance had sparked a glut of outflows across many of the older funds. This time it was the newer funds looking to sell properties quickly to remain afloat.

In 2005 and 2006, the market had moved in the funds' favour as it palmed off poor and vacant buildings to avid UK and US investors. Now, the open-ended funds faced a wasteland of buyers. Still prevented by regulation from selling for more than 5% below book value, the funds' attempts to raise cash left a trail of failed sales across Europe. Credit Suisse struggled to sell a EUR140 million office tower in Amsterdam and a EUR350 million French portfolio,[27] while in the Netherlands DEGI met dead ends for its EUR100 million portfolio in the Hague.[28] Without the sales, there was no way to generate enough money to meet redemptions so they had little choice but to stay shut. In early 2010, seven funds were still frozen.

During all of this, the German government had seen the funds stray from their mandate to be long-term, stable, income-led savings plans for the German public.[29] In 2011, it announced reforms to deaden the volatility and make the funds less appealing to short-term investors. A 24-month minimum holding period was put in place for all new investors, and 12 months for existing ones. Those redeeming less than EUR30,000 per calendar year were exempt from the rule, underscoring the message that these funds were for smaller investors.

With the passage of time and the inability to freely sell properties to improve liquidity, investors' confidence was lost and fund managers' patience was exhausted. Most frozen funds had now been closed for almost three years, the cut-off for them to decide to reopen or liquidate. With legislation now changing the very nature of the funds, many just decided not to continue.

The first to liquidate was KanAm's US fund in autumn 2010,[30] now with just six buildings left to sell,[31] followed by DEGI's EUR1.3 billion fund and Morgan Stanley's EUR2.3 billion P2 fund.[32]

Later, 15 more funds valued at EUR33 billion opted to liquidate. The majority closed in the next two years, with four holding out until 2013 and 2014. The funds from SEB and KanAm were among the largest at EUR4.8 billion and EUR3.5 billion, respectively.[33] "We finally decided to dissolve the whole fund, just to treat all the investors the same because if you reopen and the clever ones are the first ones to get out, the rest are imprisoned," said Birnbaum. Most funds were given three years to liquidate, but KanAm negotiated for closer to five years.

Knoflach also pushed hard for a five-year liquidation period for SEB's major fund. "I think it was a real tragedy for the investors in these funds because in the three-year period the market was so weak that most of the people lost a lot of their money."

The vast wave of liquidations became a concern. Forced sales would see a flood of properties equivalent to 8% of annual deal volumes hitting a market that was still in a weak state. They had already sold EUR5.7 billion of properties but in mid-2012, they still had 612 properties around to world to sell.[34] With regulators eventually allowing sales below book value, they were on course to offload another EUR18.4 billion by 2017.

It was the end of an era. From being the underestimated investors picking up gems like the Lloyd's building and moving stealthily into markets, now close to half of the industry would be dismantled.

For those that survived – Union, Deka, DWS (formerly DB Real Estate) and Commerz Real (formerly CGI) – it was size and distribution networks that had won out, but even these fund managers would require reinvention. They needed to make a determined shift from their long-standing buy-and-hold strategy to become active portfolio managers and traders. In other words, the time had come for them to be more like every other investor.

## The Lehman heart attack

In the weeks and months after the Lehman crash, the industry was moribund. The impact of Lehman's crash was immediate, a heart attack that sealed the fate of an already weakened market. European deals fell one-fifth to EUR25 billion in the third quarter of 2008. By the end of the year, deal volumes had more than halved to EUR110 billion, a figure that reflected both a fall in the number but also the lower prices at which deals were being agreed.[35] Worst hit was France, where volumes plunged 60%.

Any residual hopes for rental growth had been cruelly stubbed out. Now, 28 cities across Europe had either seen rents fall or were on the precipice. The amount of office space taken by companies in Europe dropped by one-quarter in the final three months of 2008, with the year down 12%. The fastest growing markets now went into reverse: office take-up in Dublin and Madrid both fell around 40%, and Barcelona by 23%.

Everyone retreated home as international money fell away from key markets across Europe, and debt buyers evaporated. In France, domestic institutions finished the year in a dominant position, taking up 69% of deals, almost a reversal from the year before when they represented just 35%.

The Netherlands too was back in the hands of locals, with final pre-crash deals such as a EUR650 million Corio portfolio being snapped up by Dutch investor White Estates, dodging the bullet of being sold to either Metrovacesa or Lehman.[36] Germany saw the end of major portfolio sales as international investment dropped from 66% to 51% in 2008.

The impact of the slowdown soon bled into companies' profits. Prologis posted a EUR577 million loss for the final quarter of 2008, blaming falling property values and a weakening UK pound. It was burdened by debts of EUR2 billion and was forced to sell two-thirds of its stake in its non-listed fund PEPF II at 80% below book value to raise cash.[37]

ING Real Estate recorded its first ever loss of EUR88 million for 2008 as it admitted that it had not anticipated the speed and depth of the crisis.[38]

With the slump in the number of deals, the property consultants that usually feasted on transactional fees saw profits fall through the floor. Global profits for

CBRE fell by 95% in the final quarter of 2008.[39] It later wiped off US$1.1 billion from the value of its business, a move to better reflect the value of companies it had bought recently, incurring a $1 billion loss for 2008.[40]

For the same quarter, JLL's profits fell 60% to $41.5 million as it began to shed 10% of its workforce across Europe.[41] Lay-offs were being repeated across all the agents as well as banks and fund managers; bonuses were cut, salaries frozen and graduate programmes scaled back or dropped.

The industry was being confronted with the instant repercussions of a crash in the capital markets. It was quick, undeniable and so very unlike anything it had seen before.

"I can speak from experience. The market, in effect, stopped," said Richard Plummer. "Actually it plummeted, and the ability of any property investment manager to alleviate any problems was totally frustrated by the very banks who'd lent the money. Irrespective of the quality of the assets and that of the rent-paying tenant, the banks didn't want to know and values in the sector collapsed."

This was not Paris in the early 1990s. There was no waiting for the property industry to catch up this time. Alongside all the other asset classes, property was collapsing right there in the mix. "In some ways, this cycle has to be viewed as unique because the impact was instantaneous in the banking sector," said Michael Spies. "Banks which had made commitments were really scrambling and, in some cases, lenders who had made commitments were trying to get released. Unless assets were properly capitalised they were out of luck in terms of finding liquidity."

The speed and depth of the fall in property values took the industry aback. The UK felt the full wrath of its volatile market. A dearth of transactions saw valuers send a slowing market into free fall. A 14% fall in capital values in the last quarter of 2008 meant any gains made on property in the five years to mid-2007 were now completely wiped out. For the year – the worst on record – capital values for commercial property fell 26% and total returns were -22%.[42]

Ireland's property collapse surprised few as returns slumped by 37%. France and Spain also moved into negative territory at -0.9% and -2.8%.[43] German values fell by up to 20% from their peak.[44] With data coming from largely institutional property owners, there were also concerns these figures underestimated even more dramatic falls for secondary property.

None of this fitted in with the usual pattern of events surrounding a downturn in the industry. "In normal real estate cycles, people just don't hear that the music is starting to grow a little quieter; it takes two or three years. Markets in the meantime have actually been deteriorating," said Spies. "It's the same period when usually it's the last phase of euphoria and everyone starts to develop new buildings. So oversupply kicks in as demand has waned."

This was the stage the industry believed it was at before Lehman collapsed, the liquidity on offer lulling it into a false sense of security. But this wasn't the start of a development cycle and was instead the end of a credit one.

Of course, the whole world was reeling from the unprecedented nature of the crash, its epic proportions distorting the ability to separate reality from the incredible noise. For the property industry that had dramatically changed and innovated through this period, it was a particularly unfair, and unforeseen, first test of its evolution and growth. "The thing that is common in all walks of life is that we so often make a judgement relative to our current experience," said Phil Clark, head of property at Kames Capital, in this case on how far values would fall. "Everyone says 10 to 20% valuation falls could occur. You tend not to get people saying this is going to be a 45% fall peak-to-trough and if you are in secondary shopping centres it's 70%. It's a human reaction to think in a relative way rather than a radical way, perhaps because so often the trigger for a financial crash is rarely predictable."

However, the scale of the crash could not mask the fact that with an industry so enveloped in debt, it was changing the very nature of property investing. A much larger industry meant greater lending volumes, but debt was now an integral part of a much wider spectrum of borrowers, from the opportunity funds with their mammoth debt requirements to increasingly ambitious private buyers. Then, there was the huge sector of core and value-added funds, which had converted the previously unleveraged institutional investors into regular borrowers. It was only the Germans' open-ended funds with relatively lower levels of leverage that were surviving in the debt world.

In this cycle, buyers had greater ambitions for property returns and used debt to meet these. This increased the average risk being taken across the industry even before momentum drove up prices to push that to the extreme.

This was an industry that was no longer so self-contained in its own downturn. Being attached to so much debt collectively opened the industry up to different risks in other markets, such as the chain linking securitised property debt to bulging ABS markets. "It became a financial tool, a solution, but the more debt, the more liabilities, and the more external factors, the more risk," said Pieter Hendrikse.

Property's previous incarnation as a smaller, less liquid series of domestic markets had its drawbacks. But now, by sharing a similar set-up to the rest of the financial markets, what had made property ultimately attractive – the fact that it acted differently – was now being diminished. "What was the point of real estate in the portfolio? It was not meant to be a proxy for leverage," said Noel Manns. "It was to be a less correlated asset class than the rest of the portfolio. Everyone just took it away from that."

New innovation had brought new risks. Unknowingly, in pursuing its place in the financial landscape, property had disturbed the order and appeal of its place in the portfolio. "Real estate is a phenomenally good asset over a long run for institutions to own. It should have a natural place between equities and bonds in a portfolio, but if you change the nature of the asset, you should understand that you then have a different required return. I don't think the industry really understood anything like enough of that at the time," said Peter Pereira Gray, managing partner and chief executive officer of investments for the Wellcome Trust.

Now, instead of facing oversupply from a development-led downturn, there was a devastating paralysis caused by the sudden withdrawal of liquidity, as the industry needed to get to grips with the value destruction that high levels of leverage could inflict on the market.

# 12

# BAILOUTS AND WORKOUTS

Naturally, when the time came for Songbird, the AIM-listed owner of Canary Wharf, to raise money in 2009, it did so in style.

The £880 million equity raise was the largest of any listed company in Europe that year, backed by two of the world's biggest investors. For one, the China Investment Corporation (CIC), it marked its entry onto the UK property scene.

But this was no ordinary equity raise, this was a bailout. Like the rest of the listed sector, Canary Wharf had not been immune to the crisis. In 2008, one-third of its value had been wiped off leaving it with a £2.9 billion portfolio.[1]

The timing for the capital injection was critical, just ahead of Songbird's interim results for 2009, at which point it would need to officially disclose it was in danger of breaching a £880 million loan from Citigroup. With this bailout, the world-famous business district narrowly missed going into administration for the second time in its 22-year history.[2]

The announcement was also a rare public sighting of MSREF, the Morgan Stanley fund. Noticeably, since the crash, its drumbeat of record deals and mammoth fundraising had been silenced. The only headline news was that Carrafiell – now seen as the poster child for the excesses of the bull market – had left and was advising on the Canary Wharf restructuring[3] with his new company.

The opportunity funds were missing the perfect opportunity: distress and pain, a market dislocation like never before. This was the sort of market that Morgan Stanley, Goldman Sachs and all their peers cut their teeth on, first in the US and then later in Paris. The difference now was they were the ones holding the biggest broken toys.

As the largest players of the last cycle, so would follow the greatest losses as the true impact of this style of investing played out. This would not just be the direct effect of their colossal investing programmes, but also the indirect influence that their style of higher-return investing had opened Europe up to.

US investors brought an attitude that propelled the industry into a new league, showing them that piecemeal investing could make way for much bigger and more creative deals. They also encouraged the industry up the risk spectrum, with the possibility of organised investing across a risk spectrum and through all parts of the cycle. US investors broadened the horizons of the European industry and inspired it to be more entrepreneurial on a bigger scale. The industry welcomed the chance to take part in the innovation coming across the Atlantic, whether that was CMBS or value-added investing or the increased role that debt could play in their investments.

However, embedded within all that was a culture of greater risk taking. "It encouraged [the industry] to take on more risky financial positions because they saw people like Morgan Stanley doing it and they thought it was pretty safe, and they didn't think about the risk they were taking," said one fund manager. "They simply assumed because someone else bigger and corporate was doing it, they should be doing it too."

The opportunity funds became so large that in the end they could not get their arms around their own investments, or the complexity they brought to their deals. Driven to work across different parts of the investment banks for fees, they prioritised that over the responsibilities to their fund investors, but never lost sight of their own bonuses. "They were the accelerant that made a nice warm fire into a blazing inferno in four years and it just sucked out the air," said one fund manager. "People tried to look at the fire and thought actually I can throw something into that. And it turned it into an inferno in a very short space of time."

Now, that super-smart set was about to burn up on re-entry. There was definitely a touch of schadenfreude for European onlookers but there was also a sinking feeling about what this also meant for them. The impact would reach everyone as the fallout of Europe's adoption and adaption of these ideas flowed through to workouts for CMBS, the listed market and the funds industry.

## Historic wipeouts

In early 2010, more than two and a half years after the collapse of Lehman, the news broke. Morgan Stanley admitted to investors that its US$8.8 billion MSREF VI fund would lose nearly two-thirds of its equity. The $5.4 billion loss was the largest in the history of private equity property investing.[4]

Goldman Sachs faced a similar wipeout as $1.8 billion of equity was vaporised from its Whitehall 2005 fund, leaving it down to its last $30 million.[5]

The silence from the opportunity funds until this point was an ironic role reversal back to the stubborn French banks, as they were unwilling to be disturbed from their official line on the value of their investments. "I think the reality is that the true extent of the damage was revealed quite slowly by the private equity funds," said one fund manager. "There was an effort to hold values for as long as possible in the hope that things would come back and there was an attempt to smooth the realisation of losses so that you didn't have this sudden dramatic radical markdown which was the case elsewhere."

The losses were extraordinary; the scale brought home the psychological cycle in which successful businesses can get trapped. "We are all our own worse enemies and we are all victims of our own success," said Russell Platt. "Things that we do well we tend to try to do more of, more frequently, on a bigger scale until it no longer works and if you get successful enough you blow yourself up like Morgan Stanley and Goldman Sachs."

It was irresponsible investing that lost pension funds and other clients millions and – in Morgan Stanley's case – their co-investing employees as well. The funds became a black hole of investing that rewarded fund managers handsomely on the way up but didn't take that away from them on the way down.

Morgan Stanley and Goldman Sachs were not the only ones. There were losses right across the US investors. GE Real Estate was also silent on the state of its $84 billion global real estate portfolio as shareholders questioned why it did not mark down values.[6] As the investments sat on GE's balance sheet, any problems could be kept mostly under wraps. Some analysts estimated losses were between $4 billion and $7 billion.[7] Others said if GE admitted to the losses, they were high enough to wipe out the entire conglomerate.[8]

Lehman's administrator quickly identified $15 billion of European property buried within 200 subsidiaries and joint ventures, across the UK, Sweden, France, Finland and Spain.[9] Sales started in 2009 but the administrator doubted it would claw back any of the $2.3 billion of equity owed to creditors.[10]

Unravelling Lehman's property holdings was complicated by the tangled mess across the bank's departments as it had provided a mix of debt, equity and interest rate hedging on its deals. The matrix of who was investing and how confused the most experienced of players. "Every time you met someone from Lehman, you had to spent 45 minutes figuring out which bit of the business they were from," said one fund manager.

Unlike Goldman and Morgan Stanley, Lehman did the majority of its investing using its own balance sheet, whether that was providing debt or equity or sometimes both. Its investment management was ring-fenced, as was the team in Lehman's London headquarters on 15 September 2008. "Consequently, when bankruptcy hit, we were sitting on the 29th floor and the rest of the building was being emptied," said Gerald Parkes, who ran the European real estate private equity business. This real estate business was transferred in 2009 to part of the management team and is now known as Silverpeak.

By 2010, Lehman's first fund had generated an IRR of 32% with seven assets still to sell including Dutch property company Uni-Invest, which it failed to list just before the crisis. The second fund was only generating a return of 8% and investors had been asked to inject an extra $150 million to prevent properties ending up in the hands of the banks.[11]

The public outing of MSREF's and Whitehall's losses coincided with banks putting more pressure on them to resolve their troubled assets as maturity dates for their debt loomed into view. Any chance to hold out for values to recover was still just too far out. "There was a phoney war for four or five years, everyone knew prices were low and funds were marking and operating as if prices were high. And they maintained that gap by not doing very much," said one fund manager. Their investors were happy to be part of that conspiracy rather than fully admit the substantial losses to their CIOs.

Banks had no choice. There was no way they could ignore these larger deals. Struck at high LTVs and with such poor interest coverage, banks were now calling for revaluations well ahead of the loan maturing.

One of the first to lose flight was Pegasus, MSREF's EUR2.1 billion portfolio of secondary properties bought from Union Investment. The EUR1.9 billion portfolio had 93% debt from RBS on terms which meant that the income from the properties did not cover the interest payments.[12]

The plan was for MSREF to lease the vacant space and sell assets to make up the shortfall but this was now impossible following the crash. With both sides unable to agree a price for MSREF to buy back part of the debt, and with the equity wiped out, they simply handed over the keys and RBS (in effect, the UK taxpayer) became the owner of the portfolio.

Other deals were also teetering. Whitehall's Spring portfolio, bought from DEGI for EUR2.6 billion at the top of the market, had now lost one-third of its value, wiping out the equity and the junior tranches of debt.[13] It was financed with a Credit Suisse loan sold on to 15 other banks with Deutsche Pfandbriefbank (formerly Hypo Real Estate) and French bank Nataxis taking the biggest parts. Now, just two years from maturing and with no amortisation – capital being paid down during the term of the loan – the full amount was required on deadline.

Goldman was having the same issues with the EUR1.7 billion Charlotte portfolio it had bought from Allianz at the end of 2007, with lenders RBS and HVB demanding a new valuation.[14]

Even using high levels of debt against prime properties was no help for the funds. MSREF took on the Sony Center in Berlin for a bargain price in early 2008 as part of the Japanese company's restructuring, but was still forced to sell below the asking price less than two years later.[15] In 2017, the new owner, Korean investor NPS, sold the building for EUR1.1 billion, doubling its money.[16]

## Working out in public

On the public side of the debt market, there was nowhere to hide. Even though CMBS was still no more than 10% of lending, its casualties would be the large and high-profile loans that had set it apart as a financing tool in the first place.

The workout in the CMBS market was entirely new territory. In its brief history, Europe had only ever dealt with only one default.

The fallout saw breaches of high-profile loans such as Coeur Défense and Karstadt, but the bread-and-butter workout would be for the glut of refinancing that would need to take place further down the track. With most bonds backed by loans of five to seven years, the short but rapid growth of the market meant that three-quarters of all European CMBS were due to mature between 2011 and 2014.[17]

The weakening economy would also take its toll with inevitable covenant breaches and loan defaults, leaving problem loans in the hands of special servicers, which acted for bondholders.

Also feeding through were the loans on poor-quality buildings churned out by some of the later conduits. Already in 2008, 48 tranches of CMBS had downgraded, compared with 14 in 2007.[18]

CMBS workouts were not like traditional lending, which was generally negotiations between borrower and bank. With CMBS sliced into multiple tranches, several layers of borrowers were each vying for their own best outcome.

At the high-profile end of the markets, the workouts were a masterclass in just how complex these structures had become. The US investment banks, among others, found themselves at the behest of bondholders, or protracted court proceedings as complicated rescue deals ended up setting legal and market precedents to get resolved.

The poor tenant credit that Goldman Sachs took on as it led the EUR4.5 billion Karstadt sale and leaseback came to pass in 2009 when the retailer's owner, Arcandor, filed for bankruptcy.

Goldman had been left holding its 51% stake after the credit crunch prevented its exit, with Arcandor selling the remainder to a consortium led by RREEF.

The bankruptcy left the owners of the EUR3.5 billion debt in the driving seat for the restructuring, in particular the bondholders in a EUR1.13 billion CMBS deal.

With the threat of liquidation destroying much of the value of the property and leaving them out of pocket, they agreed to extend the maturity date of the bonds in return for a 52 basis point increase on the margin paid. Until this time, the maturity of the bond had been seen as an immovable date, and this precedent needed the agreement of close to 50 bondholders as well as the senior debt and the 23 owners of the mezzanine debt.[19]

The collapse of Lehman also immediately put Coeur Défense into limbo, its future ownership uncertain as it got cornered into a three-year esoteric legal battle. The bank's collapse breached a financial covenant, enabling the bondholders to call in the EUR1.5 billion securitised loan. To avoid this, the two holding companies that owned the building went back and forth with the French courts – including to the highest court in the country – to secure "safeguarding" status, a type of bankruptcy protection usually reserved for operating companies. The holding companies won the protection, which allowed the bond to continue despite the breach, but the wrangling and uncertainty unsettled tenants as their leases expired.

As the clocked ticked down to the summer of 2014, when the bonds matured, the market braced itself for Coeur Défense's sale at a benchmark value that would damage the Paris property market.

At the last minute, Lone Star ended the saga by paying EUR1.3 billion to bondholders[20] to take control of the building. By now, Coeur Défense was ripe for an opportunity fund investor with almost one-quarter of the building empty, and at a purchase price that was around 40% less than what Lehman paid.[21]

Boutique fund manager Perella Weinberg was one of the main bondholders, buying EUR600 million of senior notes at a steep discount in 2009 from Goldman Sachs.[22] The company's managing partner, Léon Bressler, the original developer of Coeur Défense, still knew a bargain when he saw it.

## Inside the banks

While the failure of larger loans made headlines, there was no real way to get a true picture of the scale of broken property debt in Europe. With few obligations for banks to openly admit their problems, it was left to estimates and speculation as to the scale of the damage.

At the end of 2009, outstanding loans in Europe were around EUR1.84 trillion, the bulk of which was originated or refinanced near the peak of the market at LTVs of 80% plus.[23] Now, with property values falling by more than one-quarter, this potentially wiped out huge swaths of equity.[24] Those at the lower end of the LTVs may well have escaped intact, but with higher than average falls in Spain and the UK, the number of loans under water was vast. More pain was expected to come on the rest of the continent.

The anecdotal evidence from the final furlong up to the crash was also damning. "Those three or four months in Q2 of 2007, the volume of business that was written by the market in that one quarter is mind-boggling," said Peter Denton. "I've tailed back so many of the disastrous deals to that one quarter. It was indeed the last quarter before anyone realised."

A potential buyer of non-performing loans had also peeked inside some of the scarier books of the Irish banks. "I was expecting to see poor-quality collateral, aggressive lending against investment properties, but what I found across the board was equity financing dressed up as debt," they said.

Banks quickly found themselves in workout mode. First as triage centres, checking over loans with various degrees of impairment. Those with technical breaches such as breaking LTV limits that could still meet interest payments were sent to the back of the queue, making way for more pressing toxic assets with poor cashflows or those that had already defaulted on payments. In general, banks were doing anything to avoid calling in loans. Instead, "extend and pretend" was the mantra. "Like everything, you firefight. These loans are okay, let's push those forward and concentrate on the others," said one banker.

Unlike the previous cycle, the low interest rates were a help. In the previous crash, they stood at 12%, resulting in a greater number of actual rather than technical defaults. "This time around we could see the values don't look great but we still have enough interest cover and it's being serviced, can we stabilise for 12 months and just see what happens?" said the banker.

For those expecting floods of distressed properties to be sold, the banks may have appeared inactive but were being pragmatic. "Most banks did the right thing, focused on situations where good money could go after bad and there was a significant risk of their collateral value deteriorating in the hands of their borrower. Then they took action," said Richard Mully.

In some cases conversations were short as borrowers simply handed back the keys for buildings. Non-recourse lending, in which there is no comeback for the borrower other than the building, was now more prevalent, and few were used to it being enforced. "Certainly in the US non-recourse meant non-recourse and everyone knew it. If you made a non-recourse loan, you should have reflected that in your pricing," said Jeff Jacobson.

Now, German banks in particular were shocked as the lenders handed back the keys for buildings, such as with the Pegasus portfolio, with no worries about any loss of reputation. This was the very antithesis of relationship banking.

While banks reflected on how lending had become undisciplined at ground level, there were also questions as to how this had become so systemic. One major problem was the lack of oversight, with banks either not having the right risk management teams in place or those that lacked teeth in the face of the seemingly unstoppable bull run. "The risk department at the banks were the lowest paid people and were a cost centre not a revenue centre so senior management wouldn't listen to them," said a former banker. "The risk department is coming in saying we are taking way too much risk and the potential rate of default is much higher – senior management would tell them to just go away, this guy is making EUR200 million for us."

In the listed sector, banks were at the heart of mammoth recapitalisations across Spain and Germany as the global financial crisis deepened the cracks caused by the credit crunch. Immediately after the fall of Lehman, the European listed markets fell by one-fifth, the worst ever month on record, with Spain down 80% and Germany by more than half.

With a lethal combination of deteriorating share performance and falls in property values, the public scrutiny of the markets only increased the pressure on companies to address their short-term debt problems. The 32 biggest European listed companies owed around EUR23 billion in short-term debt.[25]

In Spain, Colonial and Metrovacesa moved from fire sales and internal squabbling to full-blown restructuring. Colonial put itself in the hands of the banks, securing a deal with its four main lenders – Goldman Sachs, Eurohypo, Calyon and RBS – to delay payments on EUR6.1 billion of debt for 18 months. It later sold the banks its stakes in French company SFL and Spanish building company FCC, as it posted a loss of EUR3.98 billion for 2008.[26]

Metrovacesa had already sold around EUR650 million of property to raise cash to go towards its EUR7 billion of debts. Next on the sales block was its prize property, the HSBC tower in Canary Wharf. Just 20 months after the purchase, it failed to meet its repayments for the £810 million bridge loan extended by HSBC. The UK bank bought the building back for £838 million, a 24% discount from its £1.1 billion sales price, making HSBC at least £250 million profit.[27]

Soon after, Metrovacesa ceded control of more than half of the company to six mainly Spanish banks in return for wiping out EUR2.9 billion of debt,[28] diluting the ownership of the two warring shareholders and muting their still unfinalised company split.[29]

In Germany, IVG's plan to gather up property to list in a G-REIT had failed, landing it with EUR1.9 billion of debt from its EUR3.5 billion spending spree.[30] IVG, an advocate for international investing since 1999, had been bullish on its home market in the wrong part of the cycle.

Caught up on the thrill-ride of the US money driving into the country, it had teamed up with MSREF on the Pegasus portfolio, taking 25 properties for EUR495 million and also a EUR1.3 billion portfolio of offices from Allianz.[31] It was also the new joint owner with Evans Randall of the EUR950 million building at 30 St Mary Axe, in London, known as the "Gherkin".[32] Now, under a new CEO, first order of business was to negotiate with 11 banks to extend two loans of EUR1.3 billion due to mature in 2009.[33]

## Cracks across the equity players

In the fund management industry, around EUR50 billion was wiped off the properties it managed in 2008. It was now a EUR430 billion industry, down from EUR560 billion[34] two years previously, largely a reflection of the write-down in property values across the board.

The global financial crisis was the non-listed funds' first test. It exposed a catalogue of mistakes that broke down trust with investors as they struggled to separate the consequences of the crash from the fund model.

With the property clock stopped on 15 September 2008, it was an easy reference point to identify properties that had been bought for too much and too late. "The biggest value destruction wasn't in the UK, it was in places like Spain," said Patrick Bushnell. "The much-quoted 50% value loss was rare but if you look across the industry you can find assets where people paid top dollar."

Fund performance fell to -26.8% for 2008, down from -2.5%,[35] driven by the extreme volatility of the UK market. Performance on the Continent would fall more sharply in 2009.

For many the debt bombs that detonated inside the core and value-added funds were a complete shock.

With little experience in running these financial products, fund managers faced problems that they had not tackled before. "People were property investors, they weren't money managers and they were in a crisis that they didn't know how to manage," said one fund manager. "A deal going sideways, they have an answer for. Not understanding that if you hang on [a fund] is not like a stock that when it goes to zero, it can't come back. Real estate can come back. So, it was kind of carnage there for a while."

Many properties were breaching loan covenants and fund managers had to sit between investors and banks to resolve this, requiring a different kind of skill set and a level of acceptance. "Most of the people that had come of professional age during that period had never been through a downturn," said Jacobson. "The first thing when the market cycle ended was just getting them to realise the markets turned and a lot of deals you did have gone bad and this isn't going to work out just by hoping it will. The first step is getting people to admit they were wrong whether it was because of specific mistakes, the market or both. Once you admit that things have changed and doing nothing is not an option, the next step is to make sure you have the right skills and experience to adapt to the situation. The skills needed in a workout situation are very different from the skills needed to invest and sell in a rising market."

Tishman was now regretting a German portfolio that it planned to list as a G-REIT. "We still view it as probably the best portfolio that came to market but it was 75% levered and we fought our way through it but it suffered badly through the crisis," said Michael Spies. "We thought that we had been comparatively prudent but obviously when markets turn bad 75% LTV is too much debt."

In 2008, the highest target debt levels recorded for value-added funds were as high as 82% and 80% for core. These were much more in line with an opportunistic strategy, illustrating how high levels of debt were accepted across the spectrum of risk. "It's an extraordinary thought that core funds can be 70% or 80% leverage. Of course, people were in denial even then. Core funds were still seen as core, even with high leverage," said Noel Manns.

No fund was unaffected. Even with prudent leverage, funds needed to acknowledge the dramatic step down in values across all properties, good and bad. "It was disastrous and the whole of 2009 was firefighting for most of the market, a lot of it was spent explaining to investors why we had written our investments down so dramatically – investors could not comprehend how investments could go down so much," said Gabi Stein.

Investors, which months before had fallen over themselves to get into funds, were now crushed. They felt misled, wronged even, by fund managers that lured them into what seemed like financial traps. "Investors were naïve. When it was all going well it was their decision, and all going wrong, it was the manager's decision," said Jonathan Short, who co-founded Internos Global just after the crisis.

To help resolve immediate problems such as securing refinancing, many funds needed more equity but investors were just too afraid of putting good money after bad. It was estimated that one-third of investors which were asked to commit fresh equity turned down the request.[36] "An awful lot more needed equity than received equity," said Colin Campbell. "We needed 5 million for a fund which had EUR750 million of assets and the investors wouldn't give it to us so we had to sell the fund's best property."

As well as strained relationships with fund managers, a new battle line was drawn between investors. It turned out that throwing a heterogeneous group of investors together into a single fund only remained problem-free when the market was rising and the fund was performing. Now, with problems to tackle, the large, experienced, well-resourced investors became frustrated with smaller ones who left it in their hands to resolve the fund's issues. Smaller investors, meanwhile, felt railroaded into those solutions often worked up privately between larger investors and the fund manager.

This incompatibility between investors was the single most damaging flaw for the funds, causing major ones to pull back from funds and seek much more control over their investments. They returned to the direct ownership they had sold out of 20 years before or started to invest alongside a small number of like-minded investors, squeezing out fund managers entirely or having them compete for the work with lower fees and less discretion. "No one of course knows what will happen a couple of years' time down the road," said Guido Verhoef, head of private real estate at PGGM Investments. "Then, you have to be super-flexible, and you are more flexible if you are able to have a direct say in whether you want to hold or sell. You don't want to be more or less forced to sell the best assets at the wrong price at the wrong moment in time."

# 13

## REVIVAL AND SURVIVAL

A couple of weeks before Christmas that first year after the Lehman crash, Anne Kavanagh headed to JLL's commercial property auction at a hotel in central London. The auction room was a familiar environment for Kavanagh, who worked for Lazard, taking her back to her time as a graduate at JLL.

The auctions department had been her first assignment and it was here that she had learnt about the nuts and bolts of property investing by helping to prepare lots for sale. From that, she knew that the auction room was visceral. If she needed a real gut take on the market, she would find it here.

The simple act up putting your hand up to bid for a property defied any sales tactics or prejudice. While the lots could be uninspiring high street shops or battered warehouses in provincial UK towns, this was honest pricing in a market stripped bare.

Amid the bustling room, Kavanagh was heartened. It was a small sale – just 16 lots – but there was depth to the bidding; all around hands were being raised as the hammer came down on almost 90% of the buildings at the £17 million sale that day.[1] "The man next to me was happy as he could buy his industrial unit for a 9% yield," said Kavanagh. "'Thank God', I thought to myself, 'It's still alive'." From the small scratchings of life in the auction room, Kavanagh was reassured that eventually this bustle of activity would radiate across the wider market.

Her auction buyers were right. There were good deals on offer. It might have seemed premature to talk of recovery but as early as 2009, it was there, like a hidden stream running under the markets. Mostly propelled by international money waiting in the wings and from those that had sold in the run-up to the crisis, the masters of this early recovery play would be those that could swap out debt for equity.

Kavanagh's instinct to return to the auction room was also a reflection of how the whole European property industry was feeling. Still reeling from the slam of the capital markets, its response was to retreat to the bricks and mortar. For this next phase of the cycle – in addition to sorting out the broken loans – this was an industry on a mission to deleverage and find the property again.

The industry reshaped as it unwound its debt positions. Companies failed or just fell away, including those US investment bank funds that had kick-started the boom. Fund managers tried to meet investors' new requirements for control as they rejected the funds model. Later, mergers saw listed and private players choose size as the ultimate defence in the uncertainty of a changing market and new regulation. It would be a shallow and drawn-out revival as workouts dragged on, the low interest rate environment affording banks more time to address the dead weight of loans they were carrying.

Most in the market still faced a tough few years as the European industry was characterised by inertia. With investments so badly wrecked by the crisis, people were now spending their days locked down in the dark misery of workouts. Here the property counted again as they did their best to preserve and protect value by negotiating with banks, bondholders, investors and tenants.

There was little money for new business. Listed companies were slightly ahead with money raised from rights issues. On the private side, fund managers had exciting ideas but investors were gun-shy after the shock of recent fund performance.

This was all playing out to the backdrop of potential catastrophe: Europe's sovereign debt crisis. Property activity stalled as imminent collapse of the entire European experiment – the European Union – added a new dimension of uncertainty right across the financial markets.

By 2010, there were major concerns about the levels of debt being carried by Greece, Ireland, Spain and Portugal, with the European Union and International Monetary Fund stepping in with bailouts. Economic growth in the eurozone came to a virtual standstill as all governments put in place austerity measures to cut state spending.

This only added to the resilience of the money starting to move into London from 2009 onwards. It was a testament to the depth and size of the European – and global – property industry that now even in these dire economic circumstances, there was no pause, no time for the market to fall into quiet stupor, counting the years before buyers braved the ruins.

There was now such diversity of buyers and ideas that capital was immediately on hand in all parts of the cycle. "It came back incredibly quickly because there was enough money out there on the sidelines waiting for a correction. At the top end of the market, people quickly took the opportunity to pick up really good-quality assets," said Nick Axford, who worked for CBRE.

This was a new recovery play but just for prime property. Sovereign wealth funds and major pension funds from Canada, the Middle East and Asia underpinned the market in the early period following the crash. These powerful equity-led players had the heft and scale of the Americans but appreciated property as if they were German. With debt no longer pricing them out of the market, it was like their equity counted for double while their long-term outlook and core appetite immediately made them stabilising saviours of the market. "That was the bit that saved the market because there was so much money chasing the top, that bolstered confidence," said Axford.

The wave arrived in anticipation just after the credit crunch, but picked up speed with CIC's injection of capital at Canary Wharf and NPS's purchase of the Sony Center in Berlin. In general, London was the leaping-off point for this new wave of investors as they sought trophy properties, and a bolthole away from the euro crisis.

Chinese and Korean capital was relatively new, with Middle Eastern money adding to its track record that already included the Qatari Investment Authority's backing of Songbird at Canary Wharf. Also, in a key deal for London just before the crash, four Middle Eastern banks paid less than £100 million for 80% of the development rights for the parcel of land on which the Shard was ready and waiting to be built. One investor sold its 33% stake for £30 million, losing £25 million from the book value.[2]

Moving into 2010, the trend picked up speed as Canadian pension funds invested in three properties in the City of London. The property arm of Ontario pension fund OMERS signed up to jointly develop 122 Leadenhall Street – known as the "Cheesegrater" – with British Land, reviving this building's path to the skyline.[3]

Then, one company bucking the predominant equity trend was Blackstone, paying £1.07 billion for half of the Broadgate Estate from British Land. The perceived safety of prime property saw Blackstone get pre-crash-style financing, contributing just £77 million of equity, and raising £987 million of third-party debt.[4]

This fight for core was so acute that it created a mini-bubble as the gap with secondary yields widened. Competition for limited supply for prime saw yields drop 83 basis points in six months, 35 of those in one month, the largest fall since 2003.[5] "There's a massive tension between concern over pricing versus people wanting to be in prime because they know, as and when the downturn comes, prime is what will hold its value. You need to survive a shorter period of depressed values and it will come back quicker and stronger than secondary," said Axford.

The equity buyers were having much more luck than those that had pinned their hopes on debt. As early as 2007, distressed debt funds started to raise capital, already sensing opportunity as record levels of debt spilled into ill-disciplined lending. Once this blew up, the blood would be back on the streets in a 1990s-style recovery play.

In 2010, the debt funds were still waiting. There were 34 debt funds hoping to have £20 billion of equity to invest. Instead, they had raised a respectable £12 billion but invested just £2 billion including their debt.[6]

Investors were still reticent – particularly for ideas outside the core space – or sceptical of untested concepts from fund managers. In many cases, the same managers who blew up in the crisis were now selling themselves as debt experts.

Supply was also still very low on the ground. Banks were handling the crisis more as an implosion rather than an explosion; the wave of distressed debts just never hit the streets. "Rather than being a sleepy organisation, they were acting more like principals and trying to maximise value for their shareholders by not disposing of large loan portfolios when values would have been at their lowest," said Nassar Hussain. "In the previous downturns, the commercial banks were probably a bit more sleepy and would have offloaded loans sooner only to see the buyers generate huge profits further down the line."

The funds persisted because they had grander plans than just distressed debt. They wanted to exploit the scarcity of debt across the market following the crash. For senior loans, funds were able to charge margins of 300 basis points for prime properties at LTVs of 50% to 60%.[7] Junior or mezzanine debt funds were on hand for more pressing problems, offering to fill the wide gap between lower levels of senior debt banks were prepared to lend and the equity, especially for those refinancing from top-of-market values.

The logic was right but with their early timing, they were struggling to gain traction. Tishman pulled plans for a EUR500 million mezzanine fund[8] and AREA Property Partners (formerly Apollo) put its plans for a EUR750 million dedicated debt fund on hold.[9]

It was doubly annoying for the debt funds, as they could not get ahead even though the problem was only going to get worse. On its way was a huge pipeline of refinancing barrelling towards the industry.

Around EUR480 billion of debt was due to mature across Europe in 2010 and 2011. When the borrowers came to refinance there would be a shortfall of up to EUR156 billion. This funding gap emerged as more stringent bank terms pushed strapped investors to find more equity or be unable to refinance. Banks' reluctance to lend on similar terms was made worse by the prospect of Basel III, a new revision of the capital requirements, which made it more expensive for banks to retain higher-risk loans on their balance sheets.

On hand to help was around EUR58 billion of fresh equity but this was selective, only looking at the prime end of the market. "There were still funds out there to buy sensible assets. If you were sitting on a portfolio of secondary towards tertiary regional offices, you weren't going to do well," said one banker.

The banks' pragmatism also continued with refinancing approaching. "What does the bank then do on the loan maturity?" said Jonathan Short. "Typically kick the can down the road with one-year, two-year, three-year rolling extensions."

Where the debt funds struggled, the first-generation of property-focused European opportunity funds was coming into its own. Tristan Capital Partners, Orion, AREA and Benson Elliot had all raised capital in 2008 to 2009 and were now ready to deploy. They were flexible on the opportunities: buying discounted debt or equity positions to take control of properties and portfolios. They often approached the owner of the equity to buy them out and then negotiate with the banks themselves from their stronger position. "Each is a story," said William Benjamin, head of real estate at Ares Real Estate Group. "Look at who we are buying from, why we are buying, why they are selling and how we are buying. That's not a headline grabber but every deal is a special situation where we are working it."

Ares bought a 40% stake in the Junction Unit Trust for EUR50 million, reducing the fund's debt and giving it some working capital. "A lot of it was situations
where you just had to pay down the debt. Today's equity had to replace yesterday's
debt, that was really what it was," said Benjamin.

Orion was actively buying up debt, including working with Colony Capital; the
duo bought EUR1.5 billion of Spanish company Colonial's debt at a 37% discount.[10]

Carlyle plucked a £671 million portfolio of London offices from behind a
CMBS structure in one of the biggest deals in the city in 2010.[11] Patron did the
same in a joint venture with TGP to take control of Uni-Invest, the Dutch company originally owned by Lehman. It restructured the debt to gain control of 203
secondary and tertiary properties in the Netherlands.[12]

These funds had struck a good balance, demonstrating the dexterity of the right
mix of property and financial skills. Debt was now closer to 60%. They had also
downgraded their return expectations, now looking more in the range of 15%
rather than 20% plus.[13]

Entering deals via the debt was now the best way to secure discounted opportunities. Blackstone spent more than a year amassing around 90% of Multi's debt at
discounted prices, enabling it to get control of the shopping centre developer,
which MSREF bought at the top of the cycle in 2006. Blackstone also partnered
with Canadian investment firm Ivanhoe Cambridge to buy discounted loans,
giving it a 12% stake in French REIT Gecina.[14]

Lone Star threw itself into the thorniest of the debt issues when it became the choice
of Bundesbank, the German central bank, to unpick the Excalibur CDO that Lehman
lobbed into the capital markets weeks before it crashed. This monster EUR2.9 billion
CDO was now down to 33 debt packages from 62, 20 of them in default. While it
needed patience to unravel the packages, it gave the US investor access to underlying
property including office portfolios in Germany and the Netherlands.[15]

When the loan sales from the banks finally began in 2011, they picked up at an
enormous pace; EUR220 billion were sold between 2012 and 2014. It had taken
the major banks years to get their arms around the sheer volume and complexity
of their books, as major loans sales soon became the most effective way to reduce
their balance sheets. It was a positive sign for the market as it eased the blockage
being caused by so much property still in the hands of banks. This was like the
sell-off in France in the 1990s, just with much less denial.

The buyers were mostly experienced distress players from the US buying in
bulk: Blackstone, Lone Star, Cerberus and Colony Capital.

Loan sales were centred on the UK, Ireland, Germany and Spain, and the sellers
were a reminder of the most profligate of the lenders, including RBS, Lloyds
Banking Group (formerly Lloyds TSB), Eurohypo and BBVA. Joining them were
the bailout agencies NAMA from Ireland and SAREB from Spain.

NAMA found itself continually in the spotlight as it began offloading the properties of the ambitious Irish private buyers, many of which were now in bankruptcy. It benefited from the rush for prime with around one-third of the
properties behind its loans in London. Over at Canary Wharf, NAMA received

£333 million for its share in Citigroup tower, bought by PropInvest and Derek Quinlan in the second largest UK deal, behind the HSBC tower. For £1 billion, it became another property in the hands of Middle Eastern investors.[16]

## A consolidating industry

When Short couldn't raise any money for a new fund coming out of the crisis, it was a sign that the investment management industry was changing. Well-liked and respected, Short had the reputation, the track record and the network, so raising money should have been a breeze. "We were going quite well with three capital sources until Lehman and they all pulled back," he said. "By summer 2009, we had a very loose EUR50 million from one investor, and you can't build a business around that."

Rather than give up, Short and his business partner Andrew Thornton took a different tack. Rummaging in the debris of the crisis, they pitched and won the race for GPT Halverton. An abandoned fund management business, its Australian owner was hot-footing it back down under. For EUR1, Short and Thornton took on the business and its debts, giving life to their new company Internos. "What that gave us was 80 people, five offices and EUR2 billion of assets. But all the funds were stressed and most of them were through the equity and into the debt. So, it was tough but at least we were having real conversations with people at the back end of 2009 rather than virtual ones."

The Internos story is now rare, not because of how it set up but because it was one of the very last fund managers to be able to establish a business in the new cycle. From the barriers to entry being too low, now new regulations had pulled up the drawbridge to entrants.

Soon after the crisis, the European Commission issued its plans for the Alternative Investment Fund Managers Directive (AIFMD), catching non-listed property funds in the same regulatory net as hedge funds and private equity. AIFMD regulated fund managers, monitored their activities, and required them to step up the compliance and reporting. The plus side was that once regulated, fund managers had unfettered access to market their products to investors across Europe.

The days of "two men, a dog and a basket of capital" were now long gone; costly and onerous red tape made it much more difficult for small, nimble entrepreneurial teams to break out. "In a way, it separated out the operational end from the financial end. Those people who are managing the money and others being the operating end of the business," said Kavanagh, now CIO of Patrizia.

Those looking to break out and set up new businesses now do so as operating partners or asset managers, keen to be on-the-ground experts for the larger funds allocating money. The pressure from AIFMD was felt mostly by the middle players, who needed backing to keep up with compliance and to get access to more deals and more investors. It caused a raft of buy-outs of some of the best-known pan-European entrepreneurial fund managers,.

Europa was bought by Mitsubishi Estate; Patrizia bought Rockspring; and Tristan sold 40% to Candriam Investors, an affiliate of New York Life Investment

Management.[17] After ten years, Short and Thornton sold Internos to US Principal Global Investors for between EUR40 million and EUR50 million.

With such a massive fallout from the global financial crisis, it was a surprise that listed and private companies did not start to consolidate properly until 2014.

Only one major merger took place before then, as ING Real Estate Investment Management, one of the first and largest pan-European fund managers, was put up for sale in 2009. ING Group sold it as part of a restructuring after a government bailout. CBRE bought the company in 2011 for $940 million, and merged it with its own investment management businesses to form CBRE Global Investors.[18]

Like this new platform, other mergers sought to save money through scale but also to become global players, reflecting the growing roster of international investors. In 2013, the US investment manager TIAA-CREF combined its business with Henderson Global Investors to create a £41.5 billion alliance[19] that later became a full merger as TH Real Estate in 2015.[20] In 2017, as part of a broader financial merger, the property activities of Standard Life and Aberdeen Asset Management were combined to form a £37 billion business. "Whichever industry, or whichever part of the industry you are in, everyone is looking at how many relationships can I credibly manage and shouldn't I be looking at economies of scale? We are all looking for more efficiencies," said Kavanagh.

While REITs were introduced just as the markets crashed, these new tax-efficient regimes have now helped see the listed market grow by 80%. "The REIT regime has been transformational. There is much more transparency, alignment and long-term strategies," said Harm Meijer.

Companies such as British Land, Land Securities and Unibail-Rodamco have emerged as sector or country specialists that lead in corporate governance and transparency. Consolidation has come but, like the private markets, later than expected, starting in 2014 when French company Klépierre took over Dutch company Corio to create a EUR10 billion company.[21]

Unibail-Rodamco, the largest listed European property company, and the original developer of Coeur Défense, continues to lead the ambitions of the sector. In 2018, it took over Westfield Corp., Australia's largest shopping centre operator, in a $16 billion deal, creating a $72 billion global shopping empire.[22] "In the early stage, people didn't have great faith in the long-term future of a lot of these vehicles," said Nick Jacobson. "There were quite a few which were unsustainable and just too small for the public markets. The [major players] now are the Darwinian-style survivors because they can raise capital in difficult times and so they can sustain themselves through downturns."

## Opportunity bouncing back

The death of the investment bank-led fund was understandably called quite soon after the crash. However, ten years on, they are now having a revival. "Cycles are long, memories are short," said one private equity manager.

Goldman Sachs has raised $1 billion for a new real estate fund, Broad Street Real Estate Credit Partners III, and is looking to double that to beat its $1.8 billion predecessor.[23] It has focused on debt investing in Europe and the US as it gives it more flexibility under the Volcker Rule, which now restricts US investment banks' activities following their contribution to the crash.

Morgan Stanley has also raised $2.3 billion of equity so far for its new real estate fund. Its disastrous sixth fund had returned -16% in mid-2015, not quite wiping out all investors' money as predicted in 2010.[24] Their property ambitions may not be much smaller but they are quieter, leaving room at the top for a different coterie of dominant players, all crash survivors: Blackstone, Lone Star, Starwood and Brookfield.

Blackstone and Brookfield are well ahead of the pack: Brookfield manages $155 billion[25] of property and Blackstone $120 billion.[26] Brookfield changed its name from Brascan in 2005, shortly after securing its stake in Canary Wharf. If it was a grudge for losing the first battle for the East London estate in 2004, then it was prepared to wait a long time before it took its revenge. In early 2015, along with the Qatar Investment Authority, it won a new protracted three-month battle for Canary Wharf. In the end, Songbird shareholders MSREF, Simon Glick and CIC supported the bid, valuing the company at £2.6 billion.[27]

While Blackstone was an early and consistent investor in Europe from the mid-1990s, its phenomenal growth globally has been attributed to the leadership and ambitions of Jonathan Gray, who took over as global head of real estate in 2005 and is now Blackstone's president and chief operating officer. It achieved scale without being conflicted by an advisory business, and its peers always credit it with much more of a property heart compared with banks. "When the pricing got more expensive and it became building picking and asset management, [the banks] weren't as tooled for it," said one fund manager. Blackstone's breakout deals were the $36 billion acquisition of Equity Office Properties Trust in the US in 2006,[28] followed by the $26 billion acquisition of Hilton Hotels, buying it at a premium price in 2007.

It did not go through the crisis unscathed, notably with Hilton which was written down by half before it renegotiated debt at a discount and injected in further equity to ride it through the crisis.[29] "The perception is that Blackstone walks on water and can make no fault but it vaporised equity in deals [in Europe] too but it was more rare, fewer and far between," said one fund manager.

It has also thrown its weight around this side of the Atlantic, stirring the market with the largest deal in the history of European property. Over just five years it built up Logicor, a platform of 630 logistics properties. It intended to list or sell the business, eventually attracting the deep pockets of CIC for EUR12.3 billion in the largest deal ever in the history of European real estate.[30]

With the investment banks' blow-up burnt into the collective memory of the industry, the size of Blackstone's funds and its ability to deploy that level of capital does attract criticism. In 2012, it raised $13.3 billion of equity for a global fund, and in 2017, it raised EUR7.8 billion of equity for the largest ever dedicated

European property fund. "The market has learnt nothing because Blackstone is recreating exactly the same circumstances," said one private equity manager.[31]

## Living in an artificial world

As the industry moved into 2018, a flood of liquidity was once again fuelling the market. This time it was not debt but the artificial fix of quantitative easing, money pumped into the capital markets by governments to stimulate growth after the euro crisis. It was again creating momentum as European investment volumes reached a record EUR286 billion[32] in 2017.

A decade on, and the effect of the crash has still not gone away. "We are still absolutely living with the financial crisis," said Axford. "It's still completely part of the picture because the extraordinary monetary policy that was initially put in place as an emergency measure is still dominating the capital markets. We are still living with quantitative easing and very low interest rates that are distorting price across the markets."

The industry never did go back to the simple times of the demand, supply and overdevelopment of the French property crash. The industry's life now lives on under the influence of broader macroeconomic trends on the capital markets: quantitative easing, uncertainty around the UK exit from the European Union, or "Brexit", and its contagion across Europe, and the increasingly isolationist policies of America.

The new challenge for European property is what will happen as the ECB weans the continent of this liquidity and interest rates rise, reducing property's appeal compared to other asset classes. This will be a new test for the industry. In the run-up to the crash, its appeal was underpinned by a river of debt and now it remains buoyed-up by liquidity from quantitative easing. In its 25-year modern history, it's never been in a market that's properly tested its staying power on its own merits. Today, it has become clear how much the industry has transformed in 25 years. "One of the net-net results [of the last cycle] is that property is now a truly global asset class," said Gerald Parkes. "When Teachers Insurance (TIAA) first asked me to build a strategy for Europe in 1992, I just laughed. I couldn't support that with data in most markets because if there was any, a lot of it was downright wrong."

It is strange to think about how the industry looked in the 1990s; companies were unsophisticated, markets were local and spreadsheets were novel. Now, the industry has adapted following the crash to find the right balance between financial engineering and property, and also have the right skill set to make the most of that. It's more important for graduates today to have financial acumen than property expertise, which can be learnt on the job (and that brief is now also expanding to include technology).

The level of sophistication and research to support investing programmes for national investors to global ones has changed dramatically. There is now the transparency and depth of research for investors to be confident to invest for diversification in most parts of the world rather than chasing returns. The money will ebb and flow, but fundamentally, the merits of investing in property have more recognition among the financial community that they ever did before.

The recoil from the use of debt in the crisis did jolt the industry back to the bricks and mortar. While levels of debt will vary, the crash gave the mainstream of the industry the confidence to have faith in what they were offering as property experts rather than trying to be debt players. "The good news is that much of the core investment is low leverage, not a lot of high-level leverage, so I think that's a big fundamental difference," said Michael Spies.

That said, lending is rising, this time in the hands of the alternative lenders. The debt funds eventually found favour with investors and borrowers. By mid-2015, 171 funds had closed representing over EUR100 billion of invested capital.[33]

A diversity of debt providers outside the traditional banks is considered a healthy move but with near-record highs of private debt in the market, Europe now has EUR63 billion poised to invest, five times the amount that was available in 2007.[34] It is already putting pressure on deal flow and pricing to get the capital invested.

There are concerns about having debt structured as a fund that needs to invest and exit in a defined timeframe, as well as targeting ambitious returns. "Private credit is a source of potential significant instability, and the way in which it thinks about real estate because it's not even senior. It sits behind the senior and people entering private credit funds are doing it for returns that look an awful lot like equity. Frankly, you do not get credit risk with equity style returns forever," said Simon Martin, head of research and strategy at Tristan Capital Partners.

Running alongside the financial risks this cycle are a number of secular changes that have the potential to transform the industry going forward. In this cycle, the definition of institutional property has been stretched further. Heavyweight investors, including private equity funds, have pushed into housing-linked sectors such as private rented residential, student housing and retirement housing, as well as other operating businesses such as self-storage and healthcare.

The trend is relevant to this cycle as it's about where to invest when mainstream property looks expensive. But it's also about the long term: the potential scale of these new sectors and being in step with changing demographics. These businesses deal direct with the customers so need a strong focus on customer service. This requires a shift of mindset for an industry, which has historically had a passive relationship with its "end users".

The industry is also fixated on adapting to the next generation, keen to capitalise on Millennials and Generation Z, and how they want to consume their property. It's an emphasis on shared experiences, collaborative working (and sometimes living) spaces that embrace changing technology.

Co-working space continues to be a massive trend and its broader influence on existing corporate working environments. The might of WeWork continues to astound. In just eight years, it has grown to occupy 14 million sq feet of space with lease liabilities of $18 billion over the next two decades.[35]

Its model to sub-lease that space on flexible short-term contracts to its customers (now numbering 220,000) has inspired many imitators. It also gives traditional property folk heart palpitations. They've seen this movie before: one that pitches long-term lease liabilities with fickle short-term tenants.

More worrying is the beating that retail property is taking from the rise of e-commerce. It has already contended with the decline in its traditional anchor tenants such as department stores, bookstores and electronics, but the real challenge is investing in the unknown as technology and consumer trends easily outpace lease lengths and the time and money taken to invest in new ideas. On the flip side, logistics is prospering under e-commerce, but it also has technology snapping at its heels with 3D printing, drones and automated vehicles.

Retail property's greatest defence has been the consolidation to create giants such as Unibail-Rodamco and Klépierre that can wield scale to build relationships with international retailers across their portfolios. They can also create major "destination" shopping centres to offer shopping and leisure, with an emphasis on offering the experience that doesn't come with online. Smaller centres are becoming more neighbourhood or community focused, combining shopping with services such as libraries, healthcare or co-working. Those centres in between have some thinking to do.

In, around, underneath and on top of this is the sweeping threat or opportunity of digital transformation. The term "proptech", combining property with technology, is being thrown around but the industry is really only nibbling at the edges of how technology could transform its businesses. A concrete start is how big data can help understand consumer behaviour in shopping centres or "smart" property management technology for buildings. But the real science-fiction stuff of incorporating artificial intelligence into making buildings and businesses more efficient, or how blockchain could transform trading of information and property, is still mostly the preserve of futuristic keynote speakers at conferences.

## The resilience of property

When Léon Bressler built Coeur Défense out of the rubble of a devastating property crash, his ambition was to take the best site in La Défense and develop a building to outshine all its rivals. Little did he realise that his monument to property would need the scale to survive all that was thrown at in the coming cycle.

Since it took its place on the skyline at La Défense, it has encapsulated all the excess and complexity of the times. It arrived into an emerging pan-European investing landscape before Goldman Sachs spotted the quality that made it eminently financeable, making it the classic debt play for the investment bank. Over the next two years, its value rose by EUR750 million as Lehman succumbed to the velocity of a market in momentum and made it its greatest play at the top of the cycle.

As Lehman crashed, so did Coeur Défense's value as it was dragged through courts on a nebulous legal issue, its worth slowly being dissipated by the complexity and uncertainty, as its tenants deserted it. Saved at the last minute from a fire sale that would have rocked values in the French market, 13 years on from its opening it had unnecessarily become an opportunity fund recovery play as they refurbished and re-leased it until it was fully let.

Coeur Défense's latest chapter is that it was sold by Lone Star in late 2017 for EUR1.8 billion, its exchange of ownership registering again as one of the largest single asset deals in Europe. It's now in the safer hands of Amundi Real Estate and two French institutions, Crédit Agricole Assurances and Primonial, all keen to acknowledge its symbolic status and its perfect fit for their long-term strategy. The new owners financed it with 50% LTV debt from four European banks.[36]

In all of this, Coeur Défense showed all the resilience of a good building as it met all the challenges that this cycle and global financial crisis threw at it, and like the European property industry, it's still standing.

# BIBLIOGRAPHY

## Chapter 1

1  I. Nappi-Choulet (1996). "Real Estate Crisis in the Paris Office Market", Essec Research Center, DR 96035 [online]. Available at www.essec.edu/faculty/showRef. do?bibID=2941 [Accessed 16 November 2016].
2  "Opération géante à La Défense", *Le Figaro* (4 April 1991).
3  I. Nappi-Choulet (1996). "Real Estate Crisis in the Paris Office Market", Essec Research Center, DR 96035 [online]. Available at www.essec.edu/faculty/showRef. do?bibID=2941 [Accessed 16 November 2016].
4  I. Nappi-Choulet (1996). "Real Estate Crisis in the Paris Office Market", Essec Research Center, DR 96035 [online]. Available at www.essec.edu/faculty/showRef. do?bibID=2941 [Accessed 16 November 2016].
5  I. Nappi-Choulet (1996). "Real Estate Crisis in the Paris Office Market", Essec Research Center, DR 96035 [online]. Available at www.essec.edu/faculty/showRef. do?bibID=2941 [Accessed 16 November 2016].
6  "Banks emerge from worst of Paris property recession", *EuroProperty* (February 1997), p.3.
7  "The bank that couldn't say no", *The Economist* (7 April 1994) [online]. Available at www.economist.com/node/1324442 [Accessed 16 November 2016].
8  "France offers plan to bail out Credit Lyonnais for $27 billion", *The New York Times* (18 March 1995) [online]. Available at www.nytimes.com/1995/03/18/business/ international-business-france-offers-plan-bail-credit-lyonnais-for-27-billion.html [Accessed 15 November 2016].
9  "GE Capital secures FFr 1bn French debt portfolio", *EuroProperty* (February 1997), p.3.
10 "GE Capital secures FFr 1bn French debt portfolio", *EuroProperty* (February 1997), p.3.

## Chapter 2

1  R. Hilton March (2012). "The Making of an Asset Class", *Wharton Real Estate Review* (Spring 2012) [online]. Available at http://realestate.wharton.upenn.edu/wp-content/ uploads/2017/03/726.pdf [Accessed 16 November 2016].

2   P. Linneman (2002). "The Forces Changing Real Estate Forever: Five Years Later",
    *Wharton Real Estate Review* (Fall 2002) [online]. Available at http://realestate.wharton.up
    enn.edu/working-papers/the-forces-changing-real-estate-forever-five-years-later/
    [Accessed 16 November 2016].

3   I. Nappi-Choulet (2002). "The Economic and Financial Reasons for Corporate Prop-
    erty Outsourcing in Europe", Essec Business School [online]. Available at https://
    ingridnc.files.wordpress.com/2012/12/ingridnappichoulet-ipd-2003.pdf [Accessed  16
    June 2018].

4   "Have corporate sell-offs run out of steam?", *EuroProperty* (March 2002), p.66.

5   GE Real Estate (2004). http://mba-clubs.insead.edu/private-equity/documents/GE%
    20Presentation.pdf.

6   "GE Capital builds on European foundation?", *EuroProperty* (December 1999), p.10.

7   "GE close to selling nearly all its real estate holdings", Reuters (9 April 2015) [online].
    Available  at  www.reuters.com/article/us-general-electric-divestiture/ge-close-to-selling-
    nearly-all-its-real-estate-holdings-source-idUSKBN0N020420150409 [Accessed  12  June
    2017].

8   "Morgan Stanley buys Italian property debt", *EuroProperty* (March 1997), p.1.

9   "Morgan Stanley pushes into Italy with EUR113m deal", *EuroProperty* (June 1999), p.1.

10  Beni Stabili (2000). "Telecom Italia and Lehman Brothers: New Real Estate Company
    Launched"  [online].  Available  at  www.benistabili.it/static/upload/253/253_l1_f4_
    cs-16- project-telemaco.pdf [Accessed 23 November 2016].

11  "Trio tackles Telecom Italia's assets", *EuroProperty* (December 2000/January 2001), p.1.

12  "Whitehall breaks monopoly on Swiss property holdings", *EuroProperty* (February 2001), p.7.

13  PSP Swiss Property (2001). *Annual Report*, p.9.

14  Beni Stabili (2000). "Telecom Italia and Lehman Brothers Sell Telemaco Immobiliare
    S.P.A to Whitehall" [online]. Available at www.benistabili.it/static/upload/233/233_
    l1_f4_cs-10-sale-of-telemaco-to-whitehall.pdf [Accessed 23 November 2016].

15  Beni Stabili (2000). "Telecom Italia and Lehman Brothers: New Real Estate Company
    Launched"  [online].  Available  at  www.benistabili.it/static/upload/253/253_l1_f4_
    cs-16- project-telemaco.pdf [Accessed 23 November 2016].

16  "Such a deal", *Chicago Business* (19 July 1997) [online]. Available at www.chicagobusiness.
    com/article/19970719/ISSUE01/10004694/such-a-deal [Accessed 12 June 2016].

17  "Simon Debartolo Group", *Chicago Tribune* (20 February 1998) [online]. Available at
    http://articles.chicagotribune.com/1998-02-20/business/9802200414_1_cpi-sharehol
    ders-roosevelt-field-mall-corporate-property-investors [Accessed 12 June 2016].

18  "UNIM sets the pace for Italian property", *EuroProperty* (December 1998/January
    1999), p.15.

19  "Sponda IPO outperforms", *EuroProperty* (August 1998), p.8.

20  "Rodamco Continental is top choice for investors", *EuroProperty* (August 1999), p.6.

21  Morgan Stanley Dean Witter advert, *EuroProperty* (February 2000), p.23.

22  Unibail-Rodamco (2018). "Group history" [online]. Available at www.unibail-rodam
    co.com/W/do/centre/history-performance-archive.html [Accessed 12 June 2016].

23  Morgan Stanley Dean Witter advert, *EuroProperty* (February 2000), p.23.

24  "Huge demand for Canary Wharf shares", *EG* (2 December 1999) [online]. Available
    at  www.egi.co.uk/news/huge-demand-for-canary-wharf-shares-nbsp/  [Accessed  12
    June 2016].

25  "European merger activity tops EUR11bn", *EuroProperty* (December 2000), p.10.

26  "MEPC SPECIAL: MEPC says 'enough' to the City", *EG* (2 June 2000) [online]. Available
    at  www.egi.co.uk/news/mepc-special-mepc-says-enough-to-the-city/  [Accessed
    12 June 2016].

27  "Benchmark buys in London", *EG* (2 December 1999) [online]. Available at www.egi.
    co.uk/news/benchmark-buys-in-london/ [Accessed 12 June 2016].

28  "MEPC forms joint venture with GE Capital", *EG* (11 October 1999) [online]. Available at www.egi.co.uk/news/mepc-forms-joint-venture-with-ge-capital/ [Accessed 12 June 2016].

29  "MEPC SPECIAL: MEPC says 'enough' to the City", *EG* (2 June 2002) [online]. Available at www.egi.co.uk/news/mepc-special-mepc-says-enough-to-the-city/ [Accessed 12 June 2016].

30  "Lucia", *EuroProperty* (December 2000), p.26.

31  "Deutsche launches bid for Spain's Filo", *EG* (26 May 2000) [online]. Available at www.egi.co.uk/news/deutsche-launches-bid-for-spain-s-filo/ [Accessed 12 June 2016].

32  "Silence surrounds bid for quoted UNIM," *EG* (2 December 1999) [online]. Available at www.egi.co.uk/news/silence-surrounds-bid-for-quoted-unim/ [Accessed 19 June 2018].

33  Pirelli & C. (2001) *2000 Annual Report*, pp. 23–24.

# Chapter 3

1  "Despa purchase astounds rivals", *EG* (10 February 1996) [online]. Available at www.egi.co.uk/property/home.aspx [Accessed 16 November 2016].

2  BVI, the association for investment fund management companies and investment funds (2015) Vermögen Offener Immobilienfonds.

3  "UK commercial property fall 'to beat landmark 1990s crash'", *Property Funds World* (12 December 2008) [online]. Available at www.propertyfundsworld.com/2008/12/12/uk-commercial-property-falls-beat-landmark-1990s-crash [Accessed 16 November 2016].

4  "Canary Toppled", *EG* (30 May 1992) [online]. Available at www.egi.co.uk/news/canary-toppled/ [Accessed 16 November 2016].

5  "The great credit crunch of 1992", *EG* (12 December 1992) [online]. Available at www.egi.co.uk/news/the-great-credit-crunch-of-1992/ [Accessed 16 November 2016].

6  "Germans make £72 million City buy", *EG* (15 June 1991) [online]. Available at www.egi.co.uk/news/germans-make-72m-city-buy/ [Accessed 16 November 2016].

7  "Despa secures second London investment", *EG* (2 July 1994) [online]. Available at www.egi.co.uk/news/despa-secures-second-london-investment/ [Accessed 16 November 2016].

8  "City talk suggests Despa for Postel's Hill House", *EG* (21 May 1994) [online]. Available at www.egi.co.uk/property/home.aspx [Accessed 16 November 2016].

9  "German funds: still strong on property at home and abroad", *EG* (27 August 1994) [online]. Available at www.egi.co.uk/news/german-fund-strikes-a-new-deal-in-bishopsgate/ [Accessed 16 November 2016].

10  "Despa purchase astounds rivals", *EG* (10 February 1996) [online]. Available at www.egi.co.uk/news/despa-purchase-astounds-rivals/ [Accessed 16 November 2016].

11  "Despa purchase astounds rivals", *EG* (10 February 1996) [online]. Available at www.egi.co.uk/news/despa-purchase-astounds-rivals/ [Accessed 16 November 2016].

12  "German investors turn cool on property funds", *EuroProperty* (March 1998), p.8.

13  "German investors turn cool on property funds", *EuroProperty* (March 1998), p.8.

14  "Record-breaking CGI in Dutch spending spree", *EuroProperty* (May 1997), p.4.

15  "German funds snap up top London properties", *EuroProperty* (April 1997), p.2.

16  "CGI lines up Paris office investment", *EuroProperty* (February 1998), p.1.

17  "Despa purchase astounds rivals", *EG* (10 February 1996) [online]. Available at www.egi.co.uk/news/despa-purchase-astounds-rivals/ [Accessed 16 November 2016].

18  "City developments take off", *EG* (18 February 1995) [online]. Available at www.egi.co.uk/news/city-developments-take-off/ [Accessed 16 November 2016].

19  "International buyers target London", *EuroProperty* (October 1997), p.11.

20 "The German connection", *EG* (19 April 2003) [online]. Available at www.egi.co.uk/news/the-german-connection/ [Accessed 16 November 2016].

21 "Commerz Grundbesitz Investment Gesellschaft MBH", *EuroProperty* (October 1997), p.14.

22 "German funds snap up top London properties", *EuroProperty* (April 1997), p.2.

23 "German funds halt net outflows", *EuroProperty* (November 1998), p.4.

24 "BFG makes breakthrough acquisition in Paris", *EuroProperty* (December 1997), p.5.

25 "Major projects to soak up CGI's record cashflows", *EuroProperty* (September 1999), p.3.

26 "German investors turn cool on property funds", *EuroProperty* (March 1998), p.8.

27 "German funds suffer shortfall", *EuroProperty* (November 1997), p.11.

28 BVI annual figures, author's calculation.

29 "German investors turn cool on property funds", *EuroProperty* (March 1998), p.8.

30 "Open-ended funds cast wider net", *EuroProperty* (April 1998), p.24.

31 "Open-ended funds cast wider net", *EuroProperty* (April 1998), p.24.

32 "German open end funds set to bounce back", *EuroProperty* (March 1999), p.1.

33 "German funds to extend to reach as low inflation threatens", *EuroProperty* (March 2000), p.24.

34 "German funds to extend to reach as low inflation threatens", *EuroProperty* (March 2000), p.24.

35 "German open end funds set to bounce back", *EuroProperty* (March 1999), p.1.

36 "Institutions buy into Budapest and Prague", *EuroProperty* (February 1999), p.1.

37 "HRO's sale to CGI verifies strength of Paris market", *EuroProperty* (April 1999), p.1.

38 "Paris draws EUR760m spending spree", *EuroProperty* (May 1999), p.1.

39 "Despa to aid CL in Paris", *EuroProperty* (April 2000), p.5.

40 "Strong pound tempts German investors to sell", *EG* (2 December 1999) [online]. Available at www.egi.co.uk/news/strong-pound-tempts-german-investors-to-sell/ [Accessed 16 November 2016].

41 "Nabarro Nathanson advises DEGI", *International Tax Review* (November 1998) [online]. Available at www.internationaltaxreview.com/Article/2611753/Nabarro-Nathanson-advises-DEGI.html [Accessed 16 November 2016].

42 "On their marks", *EG* (22 August 1998) [online]. Available at www.egi.co.uk/news/on-their-marks/ [Accessed 16 November 2016].

43 "German investors continue to lead as cross-border activity hits EUR19 billion", *EuroProperty* (March 2001), p.29.

# Chapter 4

1 "ProLogis secures $1.07bn in record equity drive", *EuroProperty* (October 1999), p.3.

2 INREV (2008). *Quarterly Research Report* No.17, p.3.

3 INREV (2018) *ANREV/INREV/NCREIF Fund Manager Survey 2018*, p.13.

4 "Cross border deals soar as rents recover", *EuroProperty* (March 1998), p.2.

5 "Cross border deals soar as rents recover", *EuroProperty* (March 1998), p.2.

6 "ING's asset management focus reaps rich rewards", *EuroProperty* (May 1998), p.15.

7 "ING opens real estate funds to outside investors", *EuroProperty* (December 1998/January 1999), p.2.

8 "ING raises stakes on new funds", *EuroProperty* (February 1999), p.6.

9 "Prologis acquires Kingspark Group Holdings, LTD. in the UK", Cision PR Newswire (19 August 1998) [online]. Available at www.prnewswire.co.uk/news-releases/prologis-acquires-kingspark-group-holdings-ltd-in-the-uk-156204305.html [Accessed 17 November 2016].

10 "LaSalle to buy Jones Lang Wootton for $450 million", *The New York Times* (23 October 1998) [online]. Available at www.nytimes.com/1998/10/23/business/company-news-lasalle-to-buy-jones-lang-wootton-for-450-million.html [Accessed 17 November 2016].

11  JLL (2000). "10-K" [online]. Available at http://ir.jll.com/mobile.view?c=81245&v=202&
    d=3&id=aHR0cDovL2FwaS50ZW5rd2l6YXJkLmNvbS9maWxpbmcuaG1sP2lw
    YWdlPTEzNzIzNDImRFNFUT0xJlNFUT0xMCZTUURFU0M9U0V
    DVElPTl9QQUdFJmV4cD0mc3Vic2lkPTU3 [Accessed 17 November 2016].
12  "LET revival drives EUR1bn Beckwith fund", *EuroProperty* (October 1999), p.1.
13  "Travelers backs new Orion fund", *EuroProperty* (December 1999), p.2.
14  JLL (2000). "10-K" [online]. Available at http://ir.jll.com/mobile.view?c=81245&v=202&
    d=3&id=aHR0cDovL2FwaS50ZW5rd2l6YXJkLmNvbS9maWxpbmcuaG1sP2lwYWdlP
    TEzNzIzNDImRFNFUT0xJlNFUT0xMCZTUURFU0M9U0U0VDVE          lPTl9Q
    QUdFJmV4cD0mc3Vic2lkPTU3 [Accessed 17 November 2016].

## Chapter 5

1  Unibail (2002). "Sharp rise in 2001 results" [online]. Available at: www.unibail-rodam
   co.com [Accessed 28 March 2018].
2  "Office markets show mettle in fine Euro performance", *EuroProperty* (February
   2001), p.75.
3  "Office markets show mettle in fine Euro performance", *EuroProperty* (February
   2001), p.75.
4  "Surprise standstill for European rents", *EuroProperty* (August 2001), p.1.
5  "Investment market remains strong despite 20% year-on-year slump in leasing activity",
   *EuroProperty* (February 2003), p.50.
6  "Euro Growth II fund aims to build on Euro 5 successes", *EuroProperty* (October 2001), p.4.
7  "AXA enlists German partner to launch EUR800m Paris fund", *EuroProperty* (April
   2002), p.8.
8  "Carlyle builds new EUR1.8bn fund", *EuroProperty* (August 2001), p.6.
9  "New Pricoa fund focuses on private equity strategy", *EuroProperty* (July 2001), p.2.
10 "Tishman closes pioneering European city office fund", *EuroProperty* (December 2004/
   January 2005), p.5.
11 "EUR1bn core fund marks Morgan Stanley shift", *EuroProperty* (February 2003), p.1.
12 "CDC Ixis brings Ergo into fund", *EuroProperty* (April 2002), p.3.
13 "New Heitman fund targets next wave of eastern states", *EuroProperty* (February 2003, p.5.
14 "Investment universe expands at slower pace", *EuroProperty* (August 2004), p.16.
15 "Hines nets $300m for development", *EuroProperty* (May 2003), p.7.
16 "Sonae poised to form fund for retail stakes", *EuroProperty* (October 2002), p.7.
17 "Pioneers push out from western Europe", *EuroProperty* (November 2002), pp.44–46.
18 "Sonae places 18 Iberian centres into Sierra fund", *EuroProperty* (November 2003), p.3.
19 "Ahold and AMDC form Central Europe fund", *EuroProperty* (November 2002), p.6.
20 INREV (2003) *Quarterly Research Report No.01*, p.2.
21 "East to outpace west for European returns", *EuroProperty* (March 2004), p.51.
22 "Buyers flock to the malls of central Europe", *EuroProperty* (March 2004), p.57.
23 "Two new European funds introduced by ING Real Estate Investment Management",
   *Europe Real Estate* (26 May 2004) [online]. Available at http://europe-re.com/two-new-
   european-funds-introduced-by-ing-real-estate-investment-management/36213 [Accessed
   15 November 2016].

## Chapter 6

1  DTZ (2008). *Money into Property*, p.11. Based on author's calculations from report.
2  "Terms and conditions apply", *IPE Real Assets* (December 2009) [online]. Available
   at https://realassets.ipe.com/terms-and-conditions-apply/realassets.ipe.com/terms-and-
   conditions-apply/33560.fullarticle [Accessed 20 June 2018].

3 "Mortgage bank revolution sees players define new identities", *EuroProperty* (June 2001), pp.14–17.
4 "Consolidation sweeps mortgage markets", *EuroProperty* (February 2002), pp.40–41.
5 "Aareal Bank", *EuroProperty* (December 2003/January 2004), pp.30–31.
6 "HypoVereinsbank", *EuroProperty* (August 2002), p.28.
7 "LRT wraps up Tornet acquisition in largest 2003 property deal", *EuroProperty* (February 2004), p.10.
8 "Brits bewitched at last", *EG* (13 May 2000) [online]. Available at www.egi.co.uk/ news/brits-bewitched-at-last/ [Accessed 22 February 2018].
9 "Caisse plans new La Défense bond", *EuroProperty* (July 1998), p.1.
10 "Euro investment leaps 39% as buyers ignore borders", *EuroProperty* (November 2005), pp.14, 15.
11 "UK property lending jumps £5.7bn after record quarter", *EuroProperty* (September 2004), p.6.
12 "Market worries can't muffle the boom", *EG* (5 June 2004) [online]. Available at www.egi.co.uk/property/home.aspx [Accessed 25 February 2018].
13 "Market worries can't muffle the boom", *EG* (5 June 2004) [online]. Available at www.egi.co.uk/news/market-worries-can-t-muffle-the-boom/ [Accessed 25 February 2018].
14 "German banks seek foreign break from domestic woes", *EuroProperty* (March 2005), pp.54, 55.
15 "Anglo Irish ups lending", *EG* (26 November 2005) [online]. Available at www.egi.co. uk/news/anglo-irish-ups-lending/ [Accessed 25 February 2018].
16 "UK growth drives expanding European investment market", *EuroProperty* (October 2006), p.14.
17 "Heated markets tempts lenders to lose their cool", *EuroProperty* (March 2006), p.56.
18 "New players take on cross border risks", *EuroProperty* (February 2004), pp.40, 41.

## Chapter 7

1 John Carrafiell was not available for interviews.
2 "Reality bites for the opportunity players", *EuroProperty* (July 2003), p.44.
3 "Reality bites for the opportunity players", *EuroProperty* (July 2003), p.44.
4 "Unrealistic pricing costs companies the deal", *EuroProperty* (December 2002), p.44.
5 "Pioneering Telefonica deal scuppered by price doubts", *EuroProperty* (October 2003), p.5.
6 "ENEL sells EUR1.4bn of Italian assets to consortium", *EuroProperty* (April 2004), p.4.
7 "Reality bites for the opportunity players", *EuroProperty* (July 2003), p.44.
8 A. Lin (2008). *Real Estate private equity: Market impacts on investment strategies and compositions of opportunity funds*, Master's, MIT, p.63.
9 "PSERS Commits $87.5Mln to Berwind Fund", *Commercial Real Estate Direct* (28 November 2001) [online]. Available at www.crenews.com/general_news/general/p sers- commits-87.5mln-to-berwind-fund.html [Accessed 2 February 2018].
10 "Analysts 'astonished' by Canary recommendation", *EG* (13 December 2003) [online]. Available at www.egi.co.uk/news/analysts-astonished-by-canary-recommendation/ [Accessed 2 January 2018].
11 "Morgan Stanley responds to Brascan Canary Wharf offer with raised bid", *EG* (17 January 2004) [online]. Available at www.egi.co.uk/news/morgan-stanley-responds-to-brascan-canary-wharf-offer-with-raised-bid/ [Accessed 2 March 2018].
12 Silvestor UK Properties Limited (2004). *Recommended acquisition of Canary Wharf Group PLC*, Posting of office document [online]. Available at www2.trustnet.com/Investm ents/Article.aspx?id=200401151040042682U [Accessed 2 January 2018].

13  "Put up or shut up, Canary rival told", *Evening Standard* (16 January 2004) [online]. Available at www.standard.co.uk/news/put-up-or-shut-up-canary-rival-told-6968541.html [Accessed 2 March 2018].

14  "Stakes raised in Canary bidding", *EG* (7 February 2004) [online]. Available at www.egi.co.uk/news/stakes-raised-in-canary-bidding/ [Accessed 3 March 2018].

15  "Morgan Stanley vehicle 'confident' of Canary Wharf success", *EG* (7 April 2004) [online]. Available at www.egi.co.uk/news/morgan-stanley-vehicle-confident-of-canary-wharf-success/ [Accessed 3 March 2018].

16  "Morgan's 'Songbird' bid sings for Canary", *The Telegraph* (17 April 2004) [online]. Available at www.telegraph.co.uk/finance/2883268/Morgans-Songbird-bid-sings-for-Canary.html [Accessed 3 March 2018].

17  "Reichmann sells Brascan Canary Wharf stake", *EG* (10 June 2004) [online]. Available at www.egi.co.uk/news/reichmann-sells-brascan-canary-wharf-stake/ [Accessed 3 March 2018].

18  "Landmark backers reap £1bn profit", *The Times* (11 October 2006) [online]. Available at www.thetimes.co.uk/article/landmark-backers-reap-pound1bn-profit-nvr08s2cpk5 [Accessed 2 March 2018].

19  "Fees bonanza as Morgan Stanley wins Canary", *The Times* (22 May 2004) [online]. Available at www.thetimes.co.uk/article/fees-bonanza-as-morgan-stanley-wins-canary-8ml5rsdcnv8 [Accessed 2 March 2018].

20  "Songbird completes Canary credit facilities refinancing", *EG* (8 November 2005) [online]. Available at www.egi.co.uk/news/songbird-completes-canary-credit-facilities-refinancing/ [Accessed 3 March 2018].

21  "Canary Wharf Group to pay back bonds worth £670m", *EG* (3 May 2005) [online]. Available at www.egi.co.uk/news/canary-wharf-group-to-pay-back-bonds-worth-670m/ [Accessed 3 March 2018].

22  Unibail (2004). *2004 Half Year Results*, p.14.

23  "Goldman Sachs left holding assets of struggling Karstadt", *EuroProperty* (4 February 2008), p.2.

24  "CMBS rescue deal brokered on Germany's Highstreet", *Real Estate Capital* (24 March 2010), p.12.

25  "KarstadtQuelle casts off debt burden", *Financial Times* (27 March 2006) [online]. Available at www.ft.com/content/bd885eea-bd72-11da-a998-0000779e2340 [Accessed 5 March 2018].

26  "Multi Developments", *EuroProperty* (March 2006), p.42.

27  "Multi Developments", *EuroProperty* (March 2006), p.42.

28  "Morgan Stanley becomes Europe's biggest developer", *EuroProperty* (May 2006), p.1.

29  "Morgan Stanley starts EUR2bn push into Euro hotels", *EuroProperty* (August 2006), p.1.

30  "French celebrate property tax liberation", *EuroProperty* (February 2003), pp.14–15.

31  "Metrovacesa bids EUR5.5 billion to unite Madrid and Paris", *EuroProperty* (April 2005), p.13.

32  "Colonial's French play leaves critics cold", *EuroProperty* (July 2004), p.14.

33  "Metrovacesa bids EUR5.5 billion to unite Madrid and Paris", *EuroProperty* (April 2005), p.13.

34  "… and backs Mines de la Lucette equity raising in France", *EuroProperty* (August 2006), p.3.

35  "GE closes in on Sophia takeover", *EuroProperty* (February 2004), p.4.

## Chapter 8

1  "Germany is Euro top property buyer", *EuroProperty* (April 2003), p.3.

2  "Germany is Euro top property buyer", *EuroProperty* (April 2003), p.3.

3    "Michael Koch found guilty of corruption in Frankfurt court", *IPE Real Estate* (26 March 2007) [online]. Available at https://realestate.ipe.com/michael-koch-found-guilty-of-corruption-in-frankfurt-court/23322.fullarticle [Accessed 23 July 2017].

4    "Michael Koch found guilty of corruption in Frankfurt court", *IPE Real Estate* (26 March 2007) [online]. Available at https://realestate.ipe.com/michael-koch-found-guilty-of-corruption-in-frankfurt-court/23322.fullarticle [Accessed 23 July 2017].

5    "Scope widens in German bribery investigations", *EuroProperty* (October 2004), p.1.

6    "German public turns its back on property", *EuroProperty* (April 2004), p.34.

7    CBRE (2008). *EMEA ViewPoint: German open-ended funds: past, present and future*, p.5.

8    BVI (2015). *Time series NAV and net flows of open-ended property funds_retail funds and Spezial fonds*, p.1.

9    "German public turns its back on property", *EuroProperty* (April 2004), p.34.

10   Known at this time as DGI but rebranded to DB Real Estate in 2003.

11   "DGI launches global fund", *EuroProperty* (October 2000), p.8.

12   "DEGI", *EuroProperty* (March 2004), p.40.

13   "CGI", *EuroProperty* (April 2004), p.24.

14   "KanAm carves share of record inflows", *EuroProperty* (June 2003), p.1.

15   "German funds go on EUR815m Paris spree", *EuroProperty* (August 2002), p.1.

16   "CGI seals record-breaking EUR460m Paris portfolio deal", *EuroProperty* (May 2002), p.8.

17   "iii Fonds nabs Birmingham gem", *EuroProperty* (September 2002), p.3.

18   "DB deal spree includes debut in Manchester", *EuroProperty* (September 2003), p.5.

19   "Deka buys El Triangle for EUR114.5m", *EuroProperty* (March 2002), p.5.

20   "German public turns its back on property", *EuroProperty* (April 2004), p.34.

21   "Funds suffer flipside of economic recovery", *EuroProperty* (April 2004), p.46.

22   "German public turns its back on property", *EuroProperty* (April 2004), p.34.

23   "CGI's international strategy outpaces troubled rivals", *EuroProperty* (February 2005), p.14.

24   "Banks aim to make the best of their bad lot", *EuroProperty* (February 2004), p.16.

25   "Lone Star snaps up EUR3.6bn loan portfolio from HREB", *EuroProperty* (April 2006), p.8.

26   "US fund takes on EUR1.2bn Dresdner loan portfolio", *EuroProperty* (November 2004), p.7.

27   "Downturn shows little upside for investors", *EuroProperty* (April 2003), p.14.

28   "Siemen axes EUR400m German portfolio sale", *EuroProperty* (November 2001), p.2.

29   "Residential buyers juggle politics and profits", *EuroProperty* (August 2004), p.34.

30   "Blackstone wins EUR1bn DB portfolio", *EuroProperty* (December 2002/January 2004), p.3.

31   "US funds pour capital into German resi deal", *EuroProperty* (July 2004), p.1.

32   "Residential buyers juggle politics and profit", *EuroProperty* (August 2004), p.34.

33   "Morgan Stanley joins rush to buy German housing", *EuroProperty* (February 2005), p.2.

34   "German real estate: Opportunist investors leave room for those prepared to sweat", *Euromoney* (7 December 2007) [online]. Available at www.euromoney.com/article/b1322193g10j6b/german-real-estate-opportunist-investors-leave-room-for-those-prepared-to-sweat [Accessed 27 March 2018].

35   "Viterra lands on Terra Firm in EUR7 billion homes sale", *EuroProperty* (June 2005), p.1.

36   J. Douvas, "Twenty Years of Opportunistic Real Estate Investing", *Wharton Real Estate Review* (Spring 2012).

37   "Terra Firma plans landmark Viterra securitisation deal", *EuroProperty* (September 2005), p.1.

38   "DEGI fails again to offload troubled German portfolio", *EuroProperty* (July 2005), p.1.

39   "More pain ahead for funds as failed sales halt recovery", *EuroProperty* (August 2005), p.14.

40   "DIFA EUR200m office sale marks start of German sell-off", *EuroProperty* (March 2005), p.1.

41  "Foreign sales fail to solve funds' domestic crisis", *EuroProperty* (February 2006), p.13.
42  "Bremen's space park crash lands", *EuroProperty* (October 2004), p.2.
43  "DEGI offloads two assets from failed portfolio disposal", *EuroProperty* (August 2005), p.14.
44  "DBRE write-downs secure buyer in EUR344m office deal", *EuroProperty* (December 2005/January 2006), p.14.
45  "2004: all-time record for investment", *Property Week* (17 December 2004) [online]. Available at www.propertyweek.com/news/2004-all-time-record-for-investment/3044750.article [Accessed 3 September 2017].
46  "CGI puts One Curzon Street on market", *Property Week* (17 December 2004) [online]. Available at www.propertyweek.com/news/2004-all-time-record-for-investment/3044750.article [Accessed 3 September 2017].
47  "Abu Dhabi royals splash out £520m in Mayfair", *Property Week* (15 July 2005) [online]. Available at www.propertyweek.com/news/abu-dhabi-royals-splash-out-£520m-in- mayfair/3053829.article [Accessed 3 September 2017].
48  "Deutsche Bank intervenes in frozen fund battle", *Financial News* (16 December 2005) [online]. Available at www.fnlondon.com/articles/deutsche-bank-intervenes-in-frozen- fund-battle-20051216 [Accessed 3 September 2017].
49  "Funds association steps in to tackle German crisis", *EuroProperty* (February 2006), p.1.
50  "KanAm sale raise EUR1.54bn for Grund-Invest reopening", *EuroProperty* (April 2006), p.6.
51  "DBRE boosts liquidity with Paris and Amsterdam sales", *EuroProperty* (April 2006), p.8.
52  "Spending set to top EUR200bn by year-end", *EuroProperty* (September 2006), p.7.
53  "DIY deal opens Topland's EUR1bn German spree", *EuroProperty* (June 2006), p.5.
54  "Liberty lines up German deal", *EuroProperty* (July 2006), p.8.
55  "Steep bidding sets new record yields in Frankfurt", *EuroProperty* (July 2006), p.4.
56  "German giants try to clean up portfolios with EUR3.5bn sales", *EuroProperty* (July 2006), p.3.
57  "Deka offloads EUR1.3bn assets", *EuroProperty* (October 2006), p.4.
58  "DB sells Grundbesitz portfolio at 5.3%", *EuroProperty* (15 January 2007), p.5.
59  "Goldman Sachs buys tranch[e] of Grandwert-Fonds assets", *EuroProperty* (21 May 2007), p.5.
60  "Gennies leaves Deka Immobilien after 18 months", *EuroProperty* (5 February 2007), p.1.

# Chapter 9

1  "European property market turnover hits €227bn in 2006", *IPE Real Assets* (19 March 2007) [online]. Available at https://realassets.ipe.com/european-property-market-turnover-hits-227bn-in-2006/realestate.ipe.com/european-property-market-turnover-hits-227bn-in-2006/realestate.ipe.com/european-property-market-turnover-hits-227bn-in-2006/23337.fullarticle [Accessed 11 January 2018].
2  "European investing volumes rise by more than 50% in 2006", *EuroProperty* (5 February 2007), p.3.
3  "Middle East buyers bust West End guide price", *EuroProperty* (December 2005/January 2006), p.5.
4  "Vendors see quick profit in Warsaw and Prague", *EuroProperty* (December 2006), p.3.
5  "Frankfurt and London office yields dip to record lows", *EuroProperty* (October 2006), p.3.
6  "DEGI bags Croatian Jankomir shopping mall at 5.8% yield", *EuroProperty* (August 2006), p.8.

7   "Overseas property funds far from safe as houses", *Financial Times* (14 December 2008) [online]. Available at www.ft.com/content/8f1dc4a8-ca0a-11dd-93e5-000077b07658 [Accessed 12 January 2018].

8   "DfES office turned to Irish buyer", *EG* (2 September 2016) [online]. Available at www.egi.co.uk/news/dfes-office-turned-to-irish-buyer/ [Accessed 16 November 2016].

9   "Quinlan makes Spanish debut in off-market mall deal", *EuroProperty* (June 2006), p.7.

10  INREV (2008). *Capital Raising Survey*, p.2.

11  INREV (2008). *Funds of funds study*, p.6.

12  "Alarm bells for lenders", *EuroProperty* (21 May 2007), p.14.

13  "CSFB breaks 20, refinances ELoC loan in continental CMBS", *Global Capital* (12 December 2004) [online]. Available at www.globalcapital.com/article/k58xc191035d/csfb-breaks-20-refinances-eloc-loan-in-continental-cmbs [Accessed 27 March 2018].

14  "CMBS conduits ride the rapids of demand", *Global Capital* (30 April 2007) [online]. Available at www.globalcapital.com/article/k53yrx7m13qp/cmbs-conduits-ride-the-rapids-of-demand [Accessed 27 March 2018].

15  "CMBS conduits ride the rapids of demand", *Global Capital* (30 April 2007) [online]. Available at www.globalcapital.com/article/k53yrx7m13qp/cmbs-conduits-ride-the-rapids- of-demand [Accessed 27 March 2018].

16  N. Hussain (2012). "Legacy of European CRE Lending, The Lessons Learned During The Downturn and CMBS 2.0." In: A. Petersen, ed., *Commercial Mortgage Loans and CMBS: Developments in the European Market*, 2nd ed. Sweet & Maxwell, pp.71–99.

17  "Yields have further to fall, predicts DTZ", *EuroProperty* (October 2005), p.5.

18  "Metrovacesa completes record UK deal", *EuroProperty* (7 May 2007), p.3.

19  "Metrovacesa sells back Canary Wharf HQ to HSBC", *EuroProperty* (5 December 2008) [online]. Available at www.egi.co.uk/news/metrovacesa-sells-back-canary-wharf-hq-to- hsbc/ [Accessed 13 March 2018].

20  "Union Investment sells EUR2.6 billion portfolio", *EuroProperty* (21 May 2007), p.2.

21  Union Investment (2018). *The first portfolio deals: Pegasus & Co.* [online]. Available at www.realestate.union-investment.com [Accessed 13 March 2018].

22  Lehman Brothers (2008). *Top 50 Asset Reviews*.

## Chapter 10

1   "Curzon set to turn €277m profit from German retail sale", *EG* (6 August 2007) [online]. Available at www.egi.co.uk/news/curzon-set-to-turn-277m-profit-fromgerman- retail-sale/ [Accessed 5 March 2018].

2   "Timeline: Credit crunch to downturn", BBC News (7 August 2009) [online]. Available at http://news.bbc.co.uk/1/hi/business/7521250.stm [Accessed 5 March 2018].

3   "Timeline: Credit crunch to downturn", BBC News (7 August 2009) [online]. Available at http://news.bbc.co.uk/1/hi/business/7521250.stm [Accessed 5 March 2018].

4   "CDO market seen shrinking by half in long term", Reuters (1 October 2007) [online]. Available at www.reuters.com/article/cdo-market-long-term/cdo-market-seen-shrinking- by-half-in-long-term-idUSL0192177020071001 [Accessed 14 March 2018].

5   "CMBS hits record growth but cracks are starting to show", *EG* (7 August 2007) [online]. Available at www.egi.co.uk/news/cmbs-hits-record-growth-but-cracks-are-starting- to-show/ [Accessed 6 March 2018].

6   "Securitised debt market hit by credit crisis", *EuroProperty* (3 September 2007), p.1.

7   "HSBC left with £800m hangover after record sale of its Canary Wharf tower", *The Times* (1 September 2007) [online]. Available at www.thetimes.co.uk/article/hsbc-left-with-pound800m-hangover-after-record-sale-of-its-canary-wharf-tower-gtdsjs2vlt8 [Accessed 14 March 2018].

8   "The debt spout is turned off", *EuroProperty* (5 May 2008), p.30.

9   "German banks report first-quarter profit slump", *EuroProperty* (19 May 2008), p.4.

10  "US subprime writedowns hit Eurohypo's Q2 profit", *EuroProperty* (1 September 2008), p.5.

11  "The debt spout is turned off", *EuroProperty* (5 May 2008), p.30.

12  "Money sources dry up", *EuroProperty* (4 February 2008), p.13.

13  "RBS faces bigger battle after winning fight for ABN Amro", *The Observer* (14 October 2007) [online]. Available at www.theguardian.com/business/2007/oct/14/money9 [Accessed 5 March 2018].

14  "Credit crisis shuts down Hypo's new loan business", *EuroProperty* (5 November 2007), p.3.

15  "Hypo bids EUR5.6bn for German real estate bank Depfa", *EuroProperty* (6 August 2007), p.5.

16  "JC Flowers to buy quarter of Hypo Real Estate", *Financial Times* (16 April 2008) [online]. Available at www.ft.com/content/5e9d64fe-0bc0-11dd-9840-0000779fd2ac [Accessed 5 March 2018].

17  "Banks selling off debt packages to investors", *EuroProperty* (17 December 2007), p.1.

18  "HSH Nordbank sells off EUR7.6 billion of real estate loans", *EuroProperty* (21 January 2008), p.3.

19  "Borrowers head to the mezzanine for their cash", *EuroProperty* (16 June 2008), p.9.

20  "GE shifts focus from solid assets to loan portfolios", *EuroProperty* (1 September 2008), p.1.

21  "Key data: The listed sector", *EuroProperty* (15 January 2007), p.44.

22  "REITs and wrongs", *IPE Real Assets* (September 2007) [online]. Available at https://realassets.ipe.com/reits-and-wrongs/25333.article [Accessed 10 March 2018].

23  "French SIIC 4 to target 40% float as Italy announces REITs law", *Property Finance Europe* (11 December 2006) [online]. Available at https://pie-mag.com/digitaleditions/legacy/issues/41_38%20Edition%2006-Dez-11.pdf [Accessed 10 March 2018].

24  "European real estate to get $100bn boost – S&P", *IPE Real Assets* (24 April 2006) [online]. Available at https://realassets.ipe.com/european-real-estate-to-get-100bn-boost-sp/24723.fullarticle [Accessed 10 March 2018].

25  "Unibail", *EuroProperty* (16 April 2007), p.18.

26  "ProLogis announces details of European flotation", *EG* (22 September 2006) [online]. Available at www.egi.co.uk/news/prologis-announces-details-of-european-flotation/ [Accessed 10 March 2018].

27  "Gagfah prices at top of range as European real estate IPOs surge to double last year's amount", *Global Capital* (20 October 2006) [online]. Available at www.globalcapital.com/article/k56v0h0c0dg0/gagfah-prices-at-top-of-range-as-european-real-estate-ipos-surge-to-double-last-years-amount [Accessed 10 March 2018].

28  "Alstria pioneers 'workable' German REIT legislation", *EuroProperty* (5 November 2007), p.6.

29  "Spain at risk from sub-prime", *EuroProperty* (16 April 2007), p.9.

30  "Portillo quits Colonial as share price plummets", *EuroProperty* (21 January 2008), p.6.

31  "Metrovacesa", *EuroProperty* (5 March 2007), p.15.

32  "Metrovacesa", *EuroProperty* (21 January 2008), p.24.

33  "Metrovacesa", *EuroProperty* (21 January 2008), p.24.

34  "Metrovacesa pulls out of EUR1.1bn Corio portfolio race", *EuroProperty* (7 April 2008), p.3.

35 "Slump could force sale of Metrovacesa's HSBC tower", *EuroProperty* (2 June 2008), p.5.

36 "Irish investors sidestep credit squeeze and plough on", *EuroProperty* (17 March 2008), p.15.

37 "Ashurst, Mayer Brown rethink debt package for Citigroup tower", *The Lawyer* (17 December 2007) [online]. Available at www.thelawyer.com/issues/17-december-2007/ ashurst-mayer-brown-rethink-debt-package-for-citigroup-tower/ [Accessed 5 March 2018].

38 "Deadline looms on £1bn Citigroup tower debt", *EG* (6 August 2007) [online]. Available at www.egi.co.uk/news/deadline-looms-on-1bn-citigroup-tower-debt/ [Accessed 5 March 2018].

39 "Key data: recent deals", *EG* (6 August 2007) [online]. Available at www.egi.co.uk/ news/key-data-recent-deals-76/ [Accessed 5 March 2018].

40 "Propinvest consortium buys Santander's Financial City", *EuroProperty* (4 February 2008), p.2.

41 "SEB seeks Berlin property stake sale", *Institutional Investor* (23 August 2011) [online]. Available at www.institutionalinvestor.com/article/b150z7btfsphfn/seb-seeks-berlin-property-stake-sale [Accessed 5 March 2018].

42 "The party's over: experts predict lower capital values", *EuroProperty* (1 October 2007), p.2.

43 "Funds face their biggest crisis yet", *EG* (26 January 2008) [online]. Available at www. egi.co.uk/news/funds-face-their-biggest-crisis-yet/ [Accessed 8 March 2018].

44 "Funds face their biggest crisis yet", *EG* (26 January 2008) [online]. Available at www. egi.co.uk/news/funds-face-their-biggest-crisis-yet/ [Accessed 8 March 2018].

45 "Europe clocks up best ever take-up in first half of 2007", *EuroProperty* (6 August 2007), p.6.

46 Morgan Stanley (2007). *Morgan Stanley Real Estate Raises Largest Ever Real Estate Fund with $8.0 Billion of Equity Investments* [online]. Available at www.morganstanley.com/ press-releases/morgan-stanley-real-estate-raises-largest-ever-real-estate-fund-with-80-bil lion-of-equity-investments_5075/ [Accessed 1 March 2018].

47 "MSREF fund to spend EUR6.4 billion in Europe", *EuroProperty* (April 2006), p.7.

48 "Morgan Stanley seeks to unload German assets", *EuroProperty* (3 August 2009), p.1.

49 "Morgan Stanley", *EuroProperty* (19 May 2008), p.16.

50 "Houses fall from the sky", *EuroProperty* (7 July 2008), p.14.

51 Willkie Farr & Gallagher (2008). "Frankfurt Office Advises on €3.5 Billion Sale of German Property Portfolio" [online]. Available at www.willkie.com/news/2008/06/ frankfurt-office-advises-on-35-billion-sale-of-g_ [Accessed 1 March 2018].

52 "Goldman Sachs buys German office portfolio", *EuroProperty* (21 January 2008), p.8.

53 "Goldman Sachs seeks buyers for German portfolios", *EuroProperty* (2 July 2007), p.1.

54 "Pirelli-led consortium to buy 49% stake in retail portfolio", *EuroProperty* (7 April 2008), p.8.

55 "Goldman Sachs seeks buyers for German portfolios", *EuroProperty* (2 July 2007), p.1.

56 United States Bankruptcy Court Southern District of New York (2010). "Report of Anton R. Valukas, Examiner", p.150 [online]. Available at https://jenner.com/lehman [Accessed 1 March 2018].

57 United States Bankruptcy Court Southern District of New York (2010). "Report of Anton R. Valukas, Examiner", p.365 [online]. Available at https://jenner.com/lehman [Accessed 1 March 2018].

58 United States Bankruptcy Court Southern District of New York (2010). "Report of Anton R. Valukas, Examiner", pp.104–105 [online]. Available at https://jenner.com/ lehman [Accessed 1 March 2018].

59 "Part I: the birth of Excalibur", Costar (23 February 2012) [online]. Available at https:// costarfinance.wordpress.com/2012/02/23/part-i-the-birth-of-excalibur/ [Accessed 5 March 2018].

60   "Lehman faces fight to shed real estate assets", *Financial Times* (16 August 2008) [online].
     Available at www.ft.com/content/093eb598-6af3-11dd-b613-0000779fd18c [Accessed 5
     March 2018].

## Chapter 11

1    "Hypo Real Estate chief set to quit", *Financial Times* (6 October 2008) [online]. Available
     at www.ft.com/content/6d4914f8-92bb-11dd-98b5-0000779fd18c [Accessed 12 March
     2018].
2    "Germany Agrees on New Rescue Package for HRE", *Die Spiegel* Online (6 October
     2008) [online]. Available at www.spiegel.de/international/business/response-to-financial-
     crisis-germany-agrees-on-new-rescue-package-for-hre-a-582359.html   [Accessed   12
     March 2018].
3    "Hypo Real Estate chief set to quit", *Financial Times* (6 October 2008) [online].
     Available at www.ft.com/content/6d4914f8-92bb-11dd-98b5-0000779fd18c [Acces-
     sed 12 March 2018].
4    "RBS investigation: Chapter 4: the bail out", *The Telegraph* (11 December 2011) [online].
     Available   at   www.telegraph.co.uk/finance/newsbysector/banksandfinance/8947559/
     RBS-investigation-Chapter-4-the-bail-out.html [Accessed 10 March 2018].
5    "RBS timeline: where it all went wrong", *The Telegraph* (2 December 2010) [online].
     Available   at   www.telegraph.co.uk/finance/newsbysector/banksandfinance/8176145/
     RBS-timeline-where-it-all-went-wrong.html [Accessed 10 March 2018].
6    "Deutsche Pfandbriefbank eyes EUR210bn bad loan move", *EuroProperty* (1 February
     2010), p.3.
7    "US subprime write-downs hit Eurohypo profits", *EuroProperty* (2 March 2009), p.5.
8    "S&P foresees deluge of CMBS loan defaults", *EuroProperty* (2 March 2009), p.6.
9    "Anglo Irish Bank nationalised", *The Guardian* (15 January 2009) [online]. Available at
     www.theguardian.com/business/2009/jan/15/anglo-irish-bank-nationalisation [Accessed
     12 March 2018].
10   Houses of the Oireachtas (2016). *Report of the Joint Committee of Inquiry into the Banking
     Crisis, Volume 1*, p.65 [online]. Available at https://inquiries.oireachtas.ie/banking/
     [Accessed 7 March 2018].
11   Houses of the Oireachtas (2016). *Report of the Joint Committee of Inquiry into the Banking
     Crisis, Volume 1*, pp.10–11 [online]. Available at https://inquiries.oireachtas.ie/ba
     nking/ [Accessed 7 March 2018].
12   Houses of the Oireachtas (2016). *Report of the Joint Committee of Inquiry into the Banking
     Crisis, Volume 1*, p.311 [online]. Available at https://inquiries.oireachtas.ie/banking/
     [Accessed 7 March 2018].
13   "A stopgap measure at best", *EuroProperty* (3 November 2008), p.13.
14   "A stopgap measure at best", *EuroProperty* (3 November 2008), p.13.
15   "A stopgap measure at best", *EuroProperty* (3 November 2008), p.13.
16   "Spanish caja in EUR9bn bailout", *EuroProperty* (6 April 2009), p.13.
17   "Spain's 'bad bank' Sareb doubles losses in 2014", *Financial Times* (31 March 2015)
     [online]. Available at www.ft.com/content/d0b6be40-d7b1-11e4-849b-00144feab7de
     [Accessed 15 November 2016].
18   "Sell-off by institutional investors triggers redemption closures at German open-ended prop-
     erty funds", *Property Finance Europe* (3 November 2008) [online]. Available at https:// pie-ma
     g.com/digitaleditions/legacy/issues/112_95%20PFE%20Update%2008-Nov-3.pdf [Acces-
     sed 17 March 2018].
19   "German open-ended funds", *EuroProperty* (2 June 2008), p.18.

20 "Looming slump puts German fund cashflow under strain", *EuroProperty* (3 November 2008), pp.10–11.

21 "Commerz Real joins €1bn German investment drive", *EG* (13 June 2008) [online]. Available at www.egi.co.uk/news/commerz-real-joins-1bn-german-investment-drive/ [Accessed 17 March 2018].

22 "German fund managers buy prime European assets", *EG* (15 September 2008) [online]. Available at www.egi.co.uk/news/german-fund-managers-buy-prime-europ ean-assets/ [Accessed 15 March 2018].

23 "KanAm buys OpernTurm tower from Tishman Speyer", *EG* (6 October 2008), p.2.

24 "Key data: German open-ended funds", *EuroProperty* (15 December 2008), p.35.

25 "Morgan Stanley writes off 10.4% of P2 fund value", *EuroProperty* (3 August 2009), p.2.

26 "Investors drain cash from open-ended funds in July", *EuroProperty* (21 September 2009), p.3.

27 "Funds go separate ways", *EuroProperty* (1 March 2010), p.13.

28 "DEGI seeks buyer for City of London office", *EuroProperty* (1 February 2010), p.2.

29 "Institutional exodus", *IPE Real Assets* (May 2011) [online]. Available at https://realassets. ipe.com/institutional-exodus/40644.article [Accessed 17 March 2018].

30 "KanAm liquidates US fund", *EG* (30 September 2010) [online]. Available at www. egi.co.uk/news/kanam-liquidates-us-fund/ [Accessed 20 March 2018].

31 "KanAm liquidates US fund", *EG* (30 September 2010) [online]. Available at www. egi.co.uk/news/kanam-liquidates-us-fund/ [Accessed 20 March 2018].

32 "P2 value fund closes", *EuroProperty* (1 November 2010), p.3.

33 "Open-ended funds to unload €20bn of assets", *EG* (22 October 2011) [online]. Available at www.egi.co.uk/news/open-ended-funds-to-unload-20bn-of-assets/ [Accessed 20 March 2018].

34 "Liquidation of German open-ended funds to create €20bn of opportunities", *Refire* (2 November 2012) [online]. Available at www.refire-online.com/features/investment/ liquidation-of-german-open-ended-funds-to-create-€20bn-of-opportunities/ [Accessed 20 March 2018].

35 "European real estate deal volume plummets by 55%", *EuroProperty* (19 January 2009), p.6.

36 "Local players take over", *EuroProperty* (2 February 2009), p.10.

37 "ProLogis fund management arm reports EUR577m Q4 loss", *EuroProperty* (16 February 2009), p.2.

38 "ING Real Estate records first-ever loss in 2008", *EuroProperty* (18 May 2009), p.4.

39 "CB Richard Ellis sees fourth quarter profits slump by 95%", *EuroProperty* (16 February 2009), p.3.

40 "CBRE writes down EUR1.1 billion on acquisitions as values fall", *EuroProperty* (16 March 2009), p.4.

41 "JLL European staff cull almost over", *EG* (7 February 2009) [online]. Available at www. egi.co.uk/news/jll-european-staff-cull-almost-over/ [Accessed 21 March 2018].

42 "Five years of gains wiped out by downturn, says IPD", *EG* (2 February 2009) [online]. Available at www.egi.co.uk/news/five-years-of-gains-wiped-out-by-downturn-says-ipd/ [Accessed 21 March 2018].

43 "French and Spanish property slides down the IPD index", *EuroProperty* (20 April 2009), p.4.

44 "The credit crunch finally reaches the German market", *EuroProperty* (6 October 2008), p.46.

## Chapter 12

1 "China invests in Canary Wharf with £880m bail-out of Songbird", *The Telegraph* (28 August 2009) [online]. Available at www.telegraph.co.uk/finance/newsbysector/ba

nksandfinance/6107221/China-invests-in-Canary-Wharf-with-880m-bail-out-of-Song
bird.html [Accessed 1 April 2018].

2   "China invests in Canary Wharf with £880m bail-out of Songbird", *The Telegraph* (28
    August 2009) [online]. Available at www.telegraph.co.uk/finance/newsbysector/ba
    nksandfinance/6107221/China-invests-in-Canary-Wharf-with-880m-bail-out-of-Song
    bird.html [Accessed 1 April 2018].

3   "Carrafiell firm aids big debt restructuring", *EuroProperty* (21 September 2009), p.4.

4   "Morgan Stanley Property Fund Faces $5.4 Billion Loss", *EG* (14 April 2010) [online].
    Available at www.wsj.com/articles/SB10001424052702303695604575182022093645864
    [Accessed 17 March 2018].

5   "Goldman real estate fund down to $30m", *Financial Times* (17 April 2010) [online].
    Available at www.ft.com/content/9ca7d968-48d2-11df-8af4-00144feab49a [Accessed
    17 March 2018].

6   "Big real estate write down potential overhangs GE", Reuters (17 July 2009) [online].
    Available at https://in.reuters.com/article/us-ge-realestate-analysis-sb/big-real-estate-
    writedown-potential-overhangs-ge-idINTRE56G5T720090717 [Accessed 18 March
    2018].

7   "Big real estate write down potential overhangs GE", Reuters (17 July 2009) [online]. Avail-
    able at https://in.reuters.com/article/us-ge-realestate-analysis-sb/big-real-estate-  write
    down-potential-overhangs-ge-idINTRE56G5T720090717 [Accessed 18 March 2018].

8   "Analysts suddenly notice the bomb inside GE Capital", Reuters (17 July 2009) [online].
    Available at www.businessinsider.com/henry-blodget-analysts-suddenly-worried-about-
    ges-huge-investments-in-real-estate-2009-7?IR=T [Accessed 18 March 2018].

9   "PwC aims to sell Lehman Europe assets in days", Reuters (19 September 2008)
    [online]. Available at www.reuters.com/article/us-lehman-sale/pwc-aims-to-sell-lehma
    n- europe-assets-in-days-idUSLJ42614020080919 [Accessed 18 March 2018].

10  "Lehman creditors at risk from value drop", *EuroProperty* (15 December 2008), p.5.

11  "Parties left in the cold by Lehman MBO", *EuroProperty* (17 May 2010), p.9.

12  "RBS takes over ownership of Morgan Stanley portfolio", *EuroProperty* (15 February
    2010), p.3.

13  "Spring portfolio debt holders order asset revaluation", *EuroProperty* (3 May 2010), p.1.

14  "Spring portfolio debt holders order asset revaluation", *EuroProperty* (3 May 2010), p.1.

15  "NPS completes purchase of Berlin Sony Center", Reuters (27 May 2010) [online].
    Available at www.propertyweek.com/news/nps-completes-purchase-of-berlin-sony-
    center/3163972.article [Accessed 16 March 2018].

16  "OMERS buys landmark Berlin property Sony Center for 1.1 billion euros", Reuters
    (2 October 2017) [online]. Available at https://uk.reuters.com/article/us-omers-sony/
    omers-buys-landmark-berlin-property-sony-center-for-1-1-billion-eur
    os-idUKKCN1C71DX [Accessed 16 March 2018].

17  "S&P foresees deluge of CMBS loan defaults", *EuroProperty* (2 February 2009), p.6.

18  "CMBS refinancing poses major threat to real estate market", *EuroProperty* (20 April
    2009), p.8.

19  "CMBS rescue deal brokered by Germany's Highstreet", *Real Estate Capital* (March
    2010), pp.12–13.

20  First Growth (2014). "Lone Star's swoop puts end to Windermere XII/Coeur Defense saga"
    [online]. Available at www.firstgrowthrealestatefinance.com/media_listing/lone-stars-
    swoop-puts-end-to-windermere-xiicoeur-defense-saga/ [Accessed 21 March 2018].

21  "BAML to underwrite near €1bn whole loan to finance Lone Star's Coeur Défense
    buy", *CoStar* (6 February 2014) [online]. Available at www.costar.co.uk/en/assets/
    news/2014/February/BAML-to-underwrite-near-1bn-whole-loan-to-finance-Lo
    ne-Stars- Coeur-Defense-buy/ [Accessed 20 March 2018].

22  "Lone Star in pole position to acquire Paris's Coeur Défense", *CoStar* (4 February 2014) [online]. Available at www.costar.co.uk/en/assets/news/2014/February/Lone-Star-in- pole-position-to-acquire-Pariss-Coeur-Defense/ [Accessed 20 March 2018].

23  "Banks much empty equity pot to tackle refinancing crisis", *Real Estate Capital* (April 2010), pp.8–9.

24  "Minding the debt funding gap", *Financial Times* (30 March 2010) [online]. Available at www.fit.com [Accessed 17 February 2018].

25  "Cash-strapped firms seek ways to pay back the bank", *EuroProperty* (7 April 2008), pp.10–11.

26  "Sell-offs hit Inmobiliaria Colonial's 2008 profit", *EuroProperty* (16 March 2009), p.6.

27  "HSBC buys back 8 Canada Square from Metrovacesa", *EuroProperty* (12 December 2008), p.6.

28  "Metrovacesa in debt deal", *EuroProperty* (2 February 2009), p.6.

29  "Six Creditor Banks Take Control of Metrovacesa", *The Wall Street Journal* (22 February 2009) [online]. Available at www.wsj.com/articles/SB123515263936734707 [Accessed 17 March 2018].

30  "IVG Immobilien", *EuroProperty* (17 September 2009), p.20.

31  "IVG acquires seven German offices for its planned REIT", *EuroProperty* (3 March 2007), p.4.

32  "IVG and Evans Randall buy Gherkin for EUR950m", *EuroProperty* (19 February 2007), p.4.

33  "IVG Immobilien", *EuroProperty* (17 September 2009), p.20.

34  "Consolidation looms as fund managers post falling values", *EuroProperty* (15 June 2009), p.8.

35  "INREV index reinforces economic downturn", *IPE Real Assets* (27 April 2009) [online]. Available at https://realassets.ipe.com/inrev-index-reinforces-economic-downturn/31538.fullarticle [Accessed 16 March 2018].

36  "Real estate funds adapt to a leaner environment", *EuroProperty* (21 September 2009), p.8.

## Chapter 13

1  "Bitter-sweet success for Woolworths at JLL sale", *EG* (20 December 2008) [online]. Available at www.egi.co.uk/news/bitter-sweet-success-for-woolworths-at-jll-sale/ [Accessed 3 March 2018].

2  "Qataris strike bargain deal on Shard", *EG* (26 January 2008) [online]. Available at www.egi.co.uk/news/qataris-strike-bargain-deal-on-shard/ [Accessed 13 January 2018].

3  "BL and Oxford complete Cheesegrater jv agreement", *EG* (22 December 2010) [online]. Available at www.egi.co.uk/news/bl-and-oxford-complete-cheesegrater-jv-agreement/ [Accessed 13 January 2018].

4  "British Land completes sale of half of Broadgate to Blackstone", *EG* (18 September 2009) [online]. Available at www.egi.co.uk/news/british-land-completes-sale-of-half-of-broadgate-to-blackstone/ [Accessed 17 March 2018].

5  "Yields plummet by 35 basis points during October", *EuroProperty* (16 November 2009), p.5.

6  "Investment cash eludes vulture funds", *EuroProperty* (15 February 2010), p.5.

7  INREV, CREFC, Association of Property Lenders, ZIA (2016). *Commercial Real Estate Debt in the European Economy*, p.38.

8  "Tishman Speyer pulls mezzanine debt fund", *EuroProperty* (1 June 2009), p.5.

9  "AREA's debt team leaves after fund fails to raise cash", *EuroProperty* (20 July 2009), p.3.

10  "Colony and Orion to buy Goldman debt", *EuroProperty* (7 December 2009), p.1.

11  "Opportunistic returns are hard to achieve in today's market", *EuroProperty* (2 August 2010), p.8.

12   "New Patron for Uni-Invest in European test case CMBS workout", *EG* (21 April 2012) [online]. Available at www.egi.co.uk/news/new-patron-for-uni-invest-in-european-test-case-cmbs-workout/ [Accessed 28 February 2018].

13   "Opportunistic returns are hard to achieve in today's market", *EuroProperty* (2 August 2010), p.8.

14   "Blackstone and Ivanhoe Cambridge buy loan book", *EG* (7 November 2012) [online]. Available at www.egi.co.uk/news/blackstone-and-ivanhoe-cambridge-buy-loan-book/ [Accessed 28 February 2018].

15   "Lone Star grabs Excalibur to slay Lehman's debt monster", *Real Estate Capital* (February 2012), pp.10–13.

16   "Citi Tower changes hands in £1bn deal", *The Telegraph* (25 May 2013) [online]. Available at www.telegraph.co.uk/finance/newsbysector/constructionandproperty/10080866/Citi-Tower-changes-hands-in-1bn-deal.html [Accessed 28 February 2018].

17   "London-Based Fund Managers are Selling Themselves in Droves", *Bisnow* (12 February 2018) [online]. Available at www.bisnow.com/london/news/capital-markets/london-based-fund-managers-are-selling-themselves-in-droves-84858 [Accessed 13 March 2018].

18   "TIAA-CREF and Henderson in £41.5bn tie-up", *Financial Times* (15 February 2011) [online]. Available at www.ft.com/content/55cd72c2-3919-11e0-b0f6-00144feabdc0 [Accessed 14 March 2018].

19   "TIAA-CREF and Henderson in £41.5bn tie-up", *EG* (24 June 2013) [online]. Available at www.egi.co.uk/news/tiaa-cref-and-henderson-in-41-5bn-tie-up/ [Accessed 14 March 2018].

20   "Henderson sells TH Real Estate stake to TIAA-CREF", *EG* (29 April 2015) [online]. Available at www.egi.co.uk/news/henderson-sells-th-real-estate-stake-to-tiaa-cref/ [Accessed 14 March 2018].

21   "Klépierre, Corio to merge into European retail giant – analysis", *IPE Real Assets* (29 July 2014) [online]. Available at https://realassets.ipe.com/klpierre-corio-to-merge-into-european-retail-giant-analysis/10002608.article [Accessed 21 March 2018].

22   "Westfield shareholders approve $16 bln Unibail-Rodamco deal; Chairman Lowy retires", Reuters (24 May 2018) [online]. Available at https://uk.reuters.com/article/uk-westfield-m-a-unibail-rodamco/westfield-shareholders-approve-16-bln-unibail-rodamco-deal-chairman-lowy-retires-idUKKCN1IP07L [Accessed 14 June 2018].

23   "Goldman Raised $1B for a Real Estate Fund, Putting 2007 Failure Behind It", *Bisnow* (3 July 2017) [online]. Available at www.bisnow.com/national/news/capital-markets/goldman-has-raised-1b-for-a-real-estate-fund-this-year-putting-2007-fail-behind-it-76191 [Accessed 20 March 2018].

24   "Morgan Stanley Has Raised $2.3B for a New Fund, Cementing its Real Estate Comeback", *Bisnow* (24 July 2017) [online]. Available at www.bisnow.com/national/news/capital-markets/morgan-stanley-has-raised-23b-for-a-new-fund-cementing-its-real-estate-comeback-76895? [Accessed 20 March 2018].

25   Brookfield (2018). "Real Estate" [online]. Available at www.brookfield.com/en/businesses/real-estate [Accessed 19 March 2018].

26   Blackstone (2018). "Who we are" [online]. Available at www.blackstone.com/the-firm/asset-management/real-estate [Accessed 19 March 2018].

27   "'They've seen the future and got it for a song'", *The Times* (29 January 2015) [online]. Available at www.thetimes.co.uk/article/theyve-seen-the-future-and-got-it-for-a-song-xmgx8s0bf3p [Accessed 10 March 2018].

28   Blackstone (2006). "Equity Office Agrees to be Acquired by The Blackstone Group" [online]. Available at www.blackstone.com/media/press-releases/equity-office-agrees-to-be-acquired-by-the-blackstone-group [Accessed 20 March 2018].

29  "A Surprise From Hilton: Big Profit for Blackstone", *The New York Times* (12 December 2013) [online]. Available at https://dealbook.nytimes.com/2013/12/12/a -surprise-from- hilton-big-profit-for-blackstone/ [Accessed 21 March 2018].

30  "CIC completes €12.3bn acquisition of Logicor", *EG* (30 November 2017) [online]. Available at www.egi.co.uk/news/cic-completes-e12-3bn-acquisition-of-logicor [Accessed 21 March 2018].

31  Blackstone (2017). "Blackstone Real Estate Partners Europe V Raises €7.8 Billion" [online]. Available at www.blackstone.com/media/press-releases/blackstone-real-esta te- partners-europe-v-raises-7.8-billion [Accessed 20 March 2018].

32  CBRE (2018). "2017 record year for European real estate investment" [online]. Available at http://news.cbre.eu/2017-record-year-for-european-real-estate-investm ent/ [Accessed 1 April 2018].

33  INREV, CREFC, Association of Property Lenders, ZIA (2016). *Commercial Real Estate Debt in the European Economy*, p.37.

34  Prequin (2018). *Private Debt Spotlight*, Vol. 3, Issue 3, p.4 [online]. Available at http:// docs.preqin.com/newsletters/pd/Preqin-Private-Debt-Spotlight-March-2018.pdf [Accessed 27 April 2018].

35  "WeWork's Junk Bond Adventure Raises $18 Billion Question", Bloomberg (25 April 2018) [online]. Available at www.bloomberg.com/gadfly/articles/2018-04-25/ wework- s-junk-bond-adventure-raises-18-billion-question [Accessed 27 April 2018].

36  "French investor trio confirm acquisition of Coeur Défense", *PropertyEU* (30 October 2017) [online]. Available at https://propertyeu.info/Nieuws/French-investor-trio-confirm- acqui sition-of-Coeur-Dfense/747a2a88-d015-4441-a87b-b169bc388050 [Accessed 27 April 2018].

# INDEX

9/11 terrorist attacks 50
Aareal Bank 60–61
Abbey 107
Aberdeen Asset Management 141
ABN Amro 22, 100, 107
ABP 40–41
Aegon 41
AEW Europe 48; as CDC IXIS Immo 43, 52; as Curzon Global Partners 48, 54, 104, 105
Ahold 52, 53
Ahorro Corporación Soluciones Inmobiliarias (ACSI) 119
Allianz 88, 111, 114, 128, 132
Allied Irish Bank 66, 107, 111, 118
Allied London 25
Alster, Henri 8
Alstria 110
Alternative Investment Managers Directive (AIFMD) 140
Alternative Investment Market (AIM) 75, 96, 125
AM see Multi Developments
American investors see US investors
Amstelland-MDC 53
Amundi Real Estate 146
Anglo Irish 66–67, 107, 118
AREA Property Partners 138; as Apollo Real Estate Advisors 14, 56, 109; see also Ares Management
Ares Management 138–139
Asian investors 128, 136–137
Asset-backed securities (ABS) 99–100, 124

Atemi 103
ATP 55
AXA 51, 82, 85, 119
Axford, Nick 136–137, 143

BaFin 89
BAM 78
Banco Santander 107, 111–112, 119
Bankers Trust 12, 14, 25, 61; see also Deutsche Bank
Bank of America 106, 108
Bank of England 98
Bank of Ireland 66, 107, 118
Banks: German 18, 59–60, 66, 99, 107–108, 131; Irish 66–67, 91, 99, 107, 118, 130; Spanish 118, 119, 131; UK 91, 99, 107
Barclays Capital 88, 100, 106
Barris, Roger 72
Basel II/III 66, 108, 138
Bayern LB 107
BBVA 119, 139
Belgium 22, 31, 51, 83, 86
Beni Stabili 19–20
Benjamin, William 138–139
Benson Elliot 13, 138
Beyerle, Thomas 84, 96
Birnbaum, Michael 82, 119, 121
Blackrock 100, 115
Blackstone 14, 59, 71, 86, 115, 137, 139, 142–143; Hilton Hotels 142; Logicor 142
BNP Paribas 104, 108

bonds 2, 42, 44, 101–102, 115, 124; *see also*
commercial mortgage backed securities
Borletti 78, 114
Bossom, Bruce 49
Bourdais 44
Bradford Property Trust 25
Brascan 74–75, 142; *see also* Brookfield
Breslauer, Keith 10, 49
Bressler, Léon 11, 77, 130, 145
Brexit 143
British Land 23, 64, 75, 137, 141
Broadgate 28, 64, 137
Brookfield 142
Brush, David 12, 17, 19, 70,
78, 99
Bundesbank 115, 139
Bushnell, Patrick 42, 97, 132

CBRE 44, 93, 123, 136, 141
CBRE Global Investors 141; CBRE
Investors 45
Caisse des Dépôts 18, 43, 71
Calyon 131
Campbell, Colin 45–46, 49, 54, 56, 133
Canary Wharf: battle for 69, 70, 74–76,
142; estate 4, 103, 106, 111–112, 131,
139; Group 28, 69, 78; Initial Public
Offering (IPO) 23, 109, 110;
recapitalisation 125, 137; securitisation
64, 76; Songbird Estates 75–76, 96, 125,
137, 142
Carlyle Group, The 14, 52, 139
Carrafiell, John 69, 75–76, 78, 125
Carrefour 52
Cashflow analysis 11, 15, 18, 20, 65
Castellum 21, 23
Central Europe 56, 78, 83, 95; funds
investing in 48, 52–53, 56; lending in 60,
107; transactions 36, 96
Cerberus 85–86, 88, 139
CGI: transactions 31–32, 36, 45, 80–81, 83,
89; as Commerz Real 122; development
32–34; performance 35; funds 82, 85,
120; inflows/outflows 85
China Investment Corporation (CIC) 125,
137, 142
Citigroup 69, 87–88, 125
Citigroup tower 111, 140
Clark, Phil 124
Coeur Défense; early plans for 3–5, 7, 8;
development 11, 50, 145; and Goldman
Whitehall fund 77, 145; and Lehman
Brothers 103, 106, 115, 129, 146; and
Lone Star 129, 130, 146; securitisation
106, 128

Collateralised debt obligations (CDOs)
99–101, 104–105, 107, 115, 139
Colonial 23, 79, 110–111, 131, 139
Colony Capital 25, 115, 139
commercial mortgage backed securities
(CMBS) 57–58, 67, 77, 102, 105–106,
115, 126; agented transactions 64, 69, 76;
breaches/defaults 128–129; conduit
lending 63–66, 99–101; downgrades 129;
growth of market in 66, 100, 104, 128;
and impact of credit crunch
106–107; origins in Europe 61–62; US
market 62; ratings 63, 65, 100; spreads
65, 100; workouts 126, 129, 139; *see also*
collateralised debt obligations (CDOs)
and residential mortgage-backed securities
(RMBS)
Commerz Real *see* CGI
Consortium de Realisation (CDR) 7, 12
Corio 111, 122, 141
Cornerstone 40
Corporate spin-offs 16–17, 19; energy
companies 19, 79; failed transactions 71,
76; telecoms companies 16– 20, 61, 71,
79; *see also* US investors
Corpus Sireo 20, as Corpus
Immobiliengruppe 87
Co-working 144
Credit crunch 104, 108, 110–114, 129, 131,
137; *see also* sub-prime
Crédit Agricole Assurances 146
Crédit Lyonnais 7, 36
Credit Suisse, 100, 108, 121, 128; as Credit
Suisse First Boston (CSFB) 69, 72,
73, 106
Croatia 95, 96
Curzon Global Partners *see* AEW Europe
Cycles *see* Property cycles

Data rooms 1, 5–6
DB Real Estate 82–83, 85, 88–90, 92, 122;
as DWS 122; *see also* Deutsche Bank
De Boer, Jeppe 22, 23, 110
Debt: loan to value ratios 59, 67, 105,
107–108, 112, 118, 130, 138; outstanding
volumes debt 5, 58, 66–67, 130, 144;
changing role of 6, 57–58, 68, 91, 99,
102, 124; junior 108, 11, 138; margins
60, 99, 107, 138; syndication 59, 61, 66,
106, 108; terms 59, 67; mezzanine
lending 61, 67, 99, 101,
108–109, 138; non-recourse 128, 131;
refinancing 67, 101, 107, 133, 138; senior
60–61, 67, 107–108, 138; availability of
61, 67, 98, 102, 107; workout 130, 131;

loans sales 108, 139; *see also* commercial mortgage backed securities (CMBS)
Debt investing 138–139
Debt funds 109, 137–138, 144
Degi 29, 84, 96; funds 82, 85, 119; inflows/outflows 85, 88; liquidation 121; portfolio sales 88, 92, 121, 128; transactions 28, 37, 89, 114;
Deka Immobilien; 32, 122; corruption scandal 81; as Despa 26, 28–29, 31– 36; development 36; funds 35, 82; inflows/outflows 85; performance 35; portfolio sales 91, 92; transactions 26, 28–29, 31, 33–34, 36, 84, 89, 96, 120
Denton, Peter 61, 101, 106, 130
Depfa 60, 108, 117
Despa *see* Deka
De Taurines, Christophe 73
Deutsche Annington 87, 114
Deutsche Bank 12, 60–61, 86; listed markets 25; corporate spin-offs 71; and buying retailers 78, 114, 129; as RREEF 71, 114, 129
Deutsche Bahn 87
Deutsche Pfandbriefbank 117, 128; *see also* Hypo Real Estate
Development Securities 33
Diagonal Mar 34, 96
DIFA *see* Union Investment
Diversification 35, 42, 53, 60
Doets, Jan 42
Dotcom boom/crash 24, 35, 42, 50–51, 84
Doughty Hanson 49
Dresdner Bank 60, 86
DTZ 48, 49
Duffield, John 112
DWS *see* DB Real Estate

E-commerce 145
Eichel, H. 86
Emmott, Alec 8–9
Equities 2, 13, 25, 36, 41, 44, 109, 124; *see also* listed property sector
Equity Office Properties 21, 142
Ergo 52
*Estates Gazette* 31
Établissement public pour l' aménagement de la région de la Défense (EPAD) 4,
Euro 26, 102, 110; introduction of the 16, 39, 47, 54; and convergence 101–102; membership 84; *see also* sovereign debt crisis
Eurocastle *see* Fortress

Eurohypo 60, 66–67, 99, 100, 107; losses 107; loan sales 139; transactions 88, 131
Europa Capital 48–49, 140
European Central Bank (ECB) 101, 115, 143
European Association for Investors in Non-Listed Real Estate Vehicles *see* INREV
European fund management industry *see* funds
European property markets; development 113; rental growth 39, 50, 102 113, 122; returns 50, 56, 96, 123; take up 50–51, 113; transaction volumes 39, 51, 90, 94, 122, 143; yield compression 95–96, 102, 112
European Public Real Estate Association (EPRA) 22, 79
*EuroProperty* 31, 55, 81
Eurozone 16, 67, 101, 136
Evans Randall 132
Expo Real 116, 120

FCC 111, 131
Fenk, Jürgen 60, 99
Finland 40, 79
Flight to core 136–137
Foncière des Régions 20, 79
Fonds Commun de Placement (FCP) 45–46
Fortis 107
Fortress 85, 87, 91–92, 110
France 50, 52, 78, 95, 111, 113, 122–123; 1990s crash 1–9; transactions 31, 34, 77, 83, 90, 103, 112; funds investing in 43, 47, 49, 51; debt/lending 59–60, 66; development 36, 77, 113; *see also* France Télécom, listed property sector and Real Estate Investment Trust (REITs)
France Télécom 16– 19, 79; *see also* Corporate spin-offs
Freeport 52
Friends Provident 112
Funds: alignment of interests 42, 133–134; capital raising 54, 97; core 49, 98, 124; core-plus 48, 105; debt/leverage and 47, 49, 52, 56, 59, 98, 132–133; diversification 42, 53, 55; Early models 38, 41, 42, 43; and early fund manager models 41–45, 52–53; Fonds Commun de Placement (FCP) 45–46; internal rates of return (IRR) 48, 98; investors in 54–55, 97, 133–134; limited partnerships 46; opportunity (European) 47– 49, 52, 138–139; performance 132; promote/

carry 46; recapitalisation 133; size of
51–52, 56; size of market for 39, 52, 54,
132; value add 47, 49, 55–56, 98,
124, 133;
Funds of funds 97, 120

Gagfah 87, 110
GE Capital 14, 59, 79; France Telecom
17–18; as GE Real Estate 109, 127;
MEPC 24–25; non-performing
loans 1, 8–9
Gecina 79, 139
Georgi, Richard 9, 108
Générale Continentale Investissements
(GCI) 3–4
General Motors Asset Management 74
German mortgage banks *see* banks
German open-ended funds: closures 89,
119, 120–121; corruption scandal 81, 85,
121; development 32–34, 36, 83; and
expansion outside Germany 26,
35–36, 82, 84, 96; growth of 31,
36–37, 80, 82–83, 119; inflows/outflows
34–36, 81–82, 85, 89, 92, 119–120;
investors into 34, 82, 120; liquidation of
121–122; performance 35–36, 84–85;
regulation 33, 35–36, 81, 88–89, 121;
selling 37, 88–92, 103, 122; style of
investing 27–32, 36
Germany 49, 78, 84, 95, 101, 111,
122; debt/lending 66, 139; transactions
90, 114; UK investors in 90–91;
*see also* German open-ended funds,
US investors, real estate investment
trusts, banks
Gherkin, the 132
Gilbert, Lynn 64–66, 106
Gilchrist, Robert 46
Glick, Simon 75, 142
Global financial crisis 2–3, 57, 103, 115,
131–132, 141, 146; *see also* Lehman
Brothers
Goldman Sachs, 14–15, 22–23, 72, 74,
76–77, 103, 108, 112, 125, 130–131,
142, 145; Coeur Défense 77, 145;
Whitehall fund 9, 18–19, 72–73, 75, 77,
83, 111, 126–127, 145; corporate spin
offs 18–20, 71, 75; German housing
portfolios 86, 114; German office
portfolios 92, 111, 114, 128, Karstadt
77–78, 114, 128–129
GPT Halverton 140
Granchester 45
Gray, Jonathan 142
Greece 53, 101, 136

Grosvenor 52
Grove International Partners 91, 108, 110
GSW Group 86, 88

Hammerson 52
HBOS 59, 66–67, 107
Heitman 52, 56
Henderson Global Investors 42–44, 141
Hendrikse, Pieter 41–42, 54, 97, 124
Heron International 37
Hersom, Simon 98–99, 108
Hoberg, Wenzel 32–33, 35, 81, 83
HRO International 36
HSBC 69
HSBC tower *see* Metrovacesa
HSH Nordbank 107–108
Hussain, Nassar 101, 105–106, 137–138
Hypo Real Estate 60–61, 66, 76–77, 99,
107; bailout 116–117; and Depfa merger
108, 117; non-performing loans 66, 86;
sale of stake 108
HypoVereinsbank (HVB) 114, 128

iii Fonds 83
Indirect investing 38,-39, 40, 42 *see also*
Funds
Industrial/logistics 38, 43, 49, 51, 56,
142, 145
ING Real Estate 53, 59, 93; formation 41;
funds 41, 42, 56; losses 122; sale to
CBRE 141
INREV 54
Institutional investors 2, 8, 52, 54, 97, 124;
Canadian 64, 136–137, 139; European
38, 40, 41, 48, 53, 124; US 13,-14, 38,
40, 47–49
Insurance companies *see* institutional
investors
InterContinental 78
Internos Global 133, 140–141
Investment Property Databank 54
ioGroup 51
Ireland 66, 96, 107, 118, 136, 139
Irish investors 91, 95–96, 111–112, 139
Istituto Bancario San Paulo di Torino 19
Italy 42, 49, 53, 60, 78, 79; transactions 19,
23, 86; *see also* Real Estate Investment
Trusts (REITs)
Ivanhoe Cambridge 139
IVG 51, 103, 110, 132

Jacobson, Jeff 47, 49, 51, 90, 98,
131–132
Jacobson, Nick 87, 141
JC Flowers 108, 117

JE Roberts 13, 25
Jones Lang LaSalle (JLL) 28, 37, 44, 93, 123, 135
JP Morgan 15, 17, 21, 23, 109; lending 99, 108; transactions 86, 88, 112

Kames Capital 124
KanAm Grund Group 82, 85, 89–90, 119–121
Karstadt *see* Goldman Sachs
Kavanagh, Anne 135, 140–141
Klépierre 141, 145
Knoflach, Barbara 28, 30–31, 119–121
Koch, Michael 81
KPN 40

La Défense 3–5, 11, 36, 90, 103, 145
Lahham, Aref 49
Land Securities 22, 141
LaSalle Investment Management 44–45, 47–49, 51
Lazard 135
Lee, Wilson 62–63, 101
Lehman Brothers 49, 69–70, 108–109, 111, 115, 122, 127; bankruptcy 57, 114–115, 122–123, 127; buying non-performing loans 9–10; Coeur Défense 103, 106, 145; corporate spin-offs 19–20; Excalibur 115, 139; securitisation 62–63, 106; Uni-Invest 110, 127, 139
Leverage *see* debt
Lewis, Ric 48, 51, 98, 104–105
Liberty Land 90
Linklaters 33
Linneman, Peter 71– 73, 88
Listed property sector 21–22, 64, 79, 109, 141; consolidation 24, 136, 141; equity raises/recapitalisation 125, 131, 136; France 25; initial public offering 21, 22, 23, 110; net asset value/discounts to 21, 24, 25, 71; Spain 23, 25, 79, 110, 111, 131; take privates 25, 61, 79; UK 24, 25; *see also* Real Estate Investment Trusts and Canary Wharf
Lloyds TSB 107, 139
Lloyd's building 26, 28–29, 31–32, 36, 89–90, 122
London & Regional 95
Lone Star 85–86, 129, 139, 142, 146
Lopez, Dennis 17

Manns, Noel 48, 55, 124, 133
*marchands de biens* 5
Marcus, Ian 72–73

Martin, Simon 48, 104–105, 113, 144
Maud, Glenn 111
Meijer, Harm 109–110, 141
MEPC 24–25, 65,
Merkel, Angela 90, 116, 119
Merrill Lynch 71–72, 86, 101, 114
Metro 71, 86, 90
Metrovacesa 25, 79, 111–112, 122, 131, 132; HSBC tower acquisition 103, 106, 111–112, 131
Middle Eastern investors 46, 95, 136–137, 140
Milano Centrale *see* Pirelli Real Estate
Millennials 144
Mills Corporation, The 89
Milton & Shire House 32–33
MIPIM 44, 93–94, 101, 116
Mines de la Lucette 79, 90
Mitsubishi Estates 140
Mogull, Marc 13– 15, 49, 73
Modern portfolio theory 53
Morgan, Malcolm 29
Morgan Stanley 13–15, 20, 69–70, 96, 106, 108, 114, 127, 142; Canary Wharf 23, 64, 69, 74– 76; core fund 52, 114; corporate spin-offs 71; development 78–79, 139; European Loan Conduit (ELoC) 64–65; Hotels 78–79; listed markets 21, 23, 79, 90; Real Estate Funds (MSREF) 25, 59, 69–70–74, 76, 78, 87, 103, 113–114, 125–128, 132, 139, 142; non-performing loans 19; open-ended fund 114, 119–120; Pegasus 103, 128, 132; UNIM 25, 76
Mully, Richard 91, 95, 99, 108, 110, 117, 131
Multi Developments 78, 139

Nataxis 128
National Asset Management Agency (NAMA) 118–119, 139
Nationale Nederleden (NN) 41–42
Netherlands 22, 52–53, 56, 60, 96, 99; transactions 31, 111, 121–122, 139
New Star Asset Management 112
Nielsen, Michael 55, 97
Net asset value (NAV) 21–22, 24
New York Life Investment Management 141
Non-performing loans (NPLs) 19, 21; France 1, 5–10; Germany 66, 85, 86; Ireland 118, 130; Spain 119; US 12–13, 62, 63

NPS 128, 137

O&H Property 96
Oaktree Capital 91
Offices 1, 4, 27 32, 50, 56, 70, 75, 122;
    development 33, 83; funds investing in
    41, 43, 49, 52; transactions 19, 31, 36,
    77, 84, 91, 95–96, 104, 111–112, 121,
    132, 139
OMERs 137
One Curzon Street 33, 89
O'Neill, Monica 54
Opportunity funds *see* US investors
    and *see* funds for opportunity funds
    (European)
Orion Capital Managers 49, 94,
    138–139
Orr, Robert 28, 81

Parkes, Gerald 127, 143
Patrizia 140
Patron Capital 49, 139
Pension fund *see* institutional investors
Pereira Gray, Peter 124
*pfandbriefe* 59–60, 108
PGGM 40–41, 55, 134
Perella Weinberg 130
Piani, Olivier 17–18
Pillar 52
Pirelli Real Estate 20, 78, 114; as Milano
    Centrale 19, 25, 76; as Prelios 20
Pradera 45–46, 49
Platt, Russell 13–15, 127
Plummer, Richard 44, 46, 99, 123
Potsdamer Platz 112, 114
Primonial 146
Principle Global Investors 141
Printemps 78
Private investors 96, 112, 113; *see also*
    German open-ended funds
Prologis 122; acquisitions 43, 59; fund
    38–39, 43, 45–47, 110
Property as an asset class 2–3, 13, 21, 51, 53,
    123–124, 143
Property consultants 44–45, 122, 123
Property cycles 2, 6, 9, 12, 30, 51, 91,
    123–124, 133, 136, 141
Property lending *see* debt
PropInvest 111–112, 140
Proptech 145
PSP Swiss Property Group 19

Quantitative easing 143
Quinlan, Derek 111, 140
Quinlan 96

Raingold, Paul 3–4, 7–8, 11
Raynor, James 58–59
Real Estate Investment Trusts (REITs)
    21–22, 79, 141; German REIT
    (G-REIT) 87, 109–110, 132–133; Italian
    109; as an investment exit 79, 87, 110;
    Sociétés d' Investissements Immobiliers
    Cotées (SIICs) 79, 109; UK REIT 109
Reichmann, Paul 23, 28, 69, 74
Reilly, Peter 88
Resolution Trust Corporation (RTC)
    12– 14, 62, 71, 115
Residential, private rented sector 144
Residential mortgage-backed securities
    (RMBS) 102, 105
Retail sector 145; development 32, 41,
    52–53, 80; funds 41, 43, 49, 53, 56;
    mergers and acquisitions 53, 78, 139,
    141; retail warehousing 43, 45, 49, 53;
    transactions 34, 45, 56, 65, 77–78, 84,
    90–91, 96, 112, 114,
Rinascente 78
Ritblat, John 23
Robertson, Struan 20, 72
Rockspring 99, 140; as PRICOA Property
    Investment Management 44, 46–47,
    52, 140
Rodamco 23, 53
Rodamco Europe 25, 109
Romania 95–96
Rosehaugh 28
Rothschild 77
Royal Bank of Scotland (RBS) 58–59, 66,
    76, 98, 111, 131, 139; ABN Amro
    acquisition 107; in Germany 88, 103,
    114, 128; rescue plan 117
Ruben brothers 80
Rubicon 89
Russia 66, 96, 94, 107

Savoy hotel group 111
Schaupensteiner, Wolfgang 81
Schroder Real Estate Investment
    Management 53
Scottish Equitable 112
SEB Asset Management 28–29, 35, 37,
    112, 120
Secondary properties 67, 70, 90– 92, 96,
    101, 119, 123, 137–139
Securitisation *see* commercial
    mortgage-backed securities
Securities and Exchange Commission 115
Securum 21
Shaftesbury International 83, 96
Shard, The 113, 137

Shinsei Bank 108
Short, Jonathan 133, 138, 140–141
Simon Property 21
Sinclair, Max 67
Single currency *see* euro
SITQ 64
Slade, John 32, 91
Sloane Capital 112
Slovakia 95
Sociedad de Gestión de Activos Procedentes
  de la Reestructuración Bancaria (SAREB)
  119, 139
Société Foncière Lyonnaise 8, 79
Société Générale 59, 65
Sony Center 114, 128, 137
Sovereign debt crisis 136–137, 143
Space Park, Bremen 36, 89
Spain 31, 50, 53, 56, 78, 83, 95, 101,
  107, 113, 122–123, 132; Debt/lending
  60, 118, 130, 136, 139; funds investing
  in; 49, 56; transactions 34, 43, 71, 84, 86,
  96, 112; *see also* listed property sector
Spies, Michael 102, 123, 133, 144
Sonae Inmobiliaria 53
Songbird Estates *see* Canary Wharf
Sponda 23
Standard Life 51, 141
Starwood 142
Stecher, Joseph 74
Steinbrück, Peer 119
Stein, Gabi 44, 52, 55, 97, 133
Stults, Van 49, 94
Sub-prime: lending 105, 110, 118; crisis
  104– 109, 111
Sweden 21, 40, 60, 79, 83
Switzerland 19, 61

Terra Firma 87
Thaker, Harin 99
Thornton, Andrew 140, 141
ThyssenKrupp 87
TIAA-CREF 141
Tishman Speyer 52, 102, 115, 120, 133, 138
Topland 90
Tornet 61
Tour Esso 4, 7
TGP 139
TH Real Estate 143
Transaction volumes, Europe 70, 90,
  122, 143
Tristan Capital Partners 138, 140, 144
Trausch, François 1, 5–6, 8–9, 15
Turner, Neil 53, 55, 98

UBS 96, 119
UK 43, 53, 90–91, 102, 109, 111, 113,
  136; 1990s crash 27–29, 44; Brexit 143;
  Debt/lending 60, 66, 99, 130, 139;
  development 32–33, 78, 80, 113;
  performance 112, 119, 123, 132; take up
  50, 70; transactions 26, 28–29, 31, 37, 83,
  95–96, 119–120, 132, 137, 139
Unibail-Rodamco 141, 145; Coeur
  Défense 11, 50, 77; Rodamco Europe
  acquisition 109; securitisation 77;
  as Unibail 23, 113
Uni-Invest *see* Lehman Brothers
UNIM 23, 25, 76
Union Investment 103, 120, 122, 128; as
  DIFA 28, 33, 35, 88–89
Urban Land Institute 8
US investors; corporate spin-offs 16–20, 71,
  86; German housing sales 86– 88, 110;
  German open-ended fund acquisitions
  91–92, 103; listed sector/REITS 20–25,
  79; losses 126–128; non-performing loans
  5–8, 9–10; opportunity funds 70–73, 79,
  87 95, 114, 125–126, 141–143; returns
  70, 72–73, 77, 87, 127; style of investing
  6, 8–15, 126; *see also*
  institutional investors
US savings and loans crisis 12, 21

Valente, Joe 49, 68, 102
van Dijkum, Floris 21–23, 25
Verhoef, Guido 134
Vesteda 40
Viterra 87–88
Vivendi Group 23
Vrensen, Hans 57

Wang, Nina 80
Weinberg, John 23
Wellcome Trust 124
Westfield 65, 80, 141
WestInvest 85
WeWork 144
Whight, Paul 45
White City 80
Whitehall fund *see* Goldman Sachs
  Whitehall fund
Wilkinson, Rob 43, 48,
  56, 101
Workouts 136

Zehner, Jon 21–22, 24
Zell, Sam 21–22

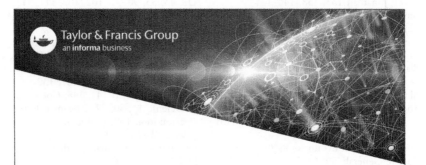